T4-AHE-474

Lecture Notes in Economics and Mathematical Systems   582

Jochen Andritzky

Sovereign Default Risk Valuation

Implications of Debt Crises and Bond Restructurings

With 43 Figures
and 49 Tables

 Springer

Jochen Andritzky
6432 Divine St
McLean, VA 22101
USA
jochen@andritzky.com

336.34
A57a

ISBN-10 3-540-37448-5   Springer Berlin Heidelberg New York
ISBN-13 978-3-540-37448-0   Springer Berlin Heidelberg New York

Springer is a part of Springer Science+Business Media
springeronline.com

© Springer Berlin Heidelberg 2006

Typesetting: Camera ready by author
Cover: Erich Kirchner, Heidelberg
Production: LE-TEX, Jelonek, Schmidt & Vöckler GbR, Leipzig

SPIN 11818731   Printed on acid-free paper – 88/3100 – 5 4 3 2 1 0

Preface

When politics meet economics, history has taught that standard concepts of welfare maximization and efficiency take a back seat. And if financial theory by itself were to guide investment decisions, empirical finance could not produce new findings. Combine politics and economics, theoretical and empirical finance, and add some spice from dynamically developing markets, and you end up valuing emerging market sovereign bonds. The monograph at hand approaches this challenge. I am grateful for all the support and advice that I received during this ambitious project.

First let me thank my advisors at the University of St. Gallen, Klaus Spremann and Paul Söderlind, who offered me the opportunity to join the Swiss Institute for Banking and Finance (s/bf) for fruitful years of research and who supported my academic progress. I also thank Kenneth Kletzer and Michael Dooley of the University of California, Santa Cruz, for hosting me as a visiting scholar and providing straightforward advice on my work. My research abroad was made possible through the sponsorship of the Swiss National Fund, for which I am most grateful.

As with most academic studies, this monograph has benefitted greatly from the comments of others. Foremost, I thank Manmohan Singh from the International Monetary Fund (IMF) whose good sense of financial markets provided a refreshing antidote to my predominantly theoretical perspective. With his input and his extensive contacts among market insiders two joint working papers came into being, thus establishing the basis for Chap. 6. Furthermore, I thank Bojan Bistrovic for his thoughts on many mathematical issues.

A summer spent at the IMF in 2004 became a very fruitful catalyst for my research. I would like to thank everyone who made this possible. The resulting IMF working paper constitutes an excellent complement to this monograph and helped me gauge the fundamental drivers of sovereign bond risk. Hence I owe a debt of gratitude to my co-authors, Natalia Tamirisa from the Policy Development and Review Department and Geoffrey Bannister from the International Capital Markets Department, as well as to my advisor Martine Guerguil. Many contacts from that time have provided helpful guidance and have given rise to important discussions transcending the scope of this dissertation. For their inspiration I would like to thank Axel Bertuch-Samuels, Peter Breuer, Jorge Chan-Lau, Norbert Funke, and Alexander Plekhanov. Frank Packer and Haibin Zhu from the Bank for International Settlements contributed valuable comments to Chap. 6. Raúl Javaloyes from the UNCTAD gave me helpful insight into debt management tools. Alvin Ying from JP Morgan Chase always responded instantly to data requests.

Profound thanks, too, are due to my colleagues at my home university in Switzerland who not only made the time there so enjoyable but also provided crucial support in the early stages of this project. Helpful comments were also received from the conferences of the German Economic Association, the Swiss Society for Financial Market Research, the Irish Economic Association, and the Quant Congress USA, as well as from seminar participants at the universities in St. Gallen and Santa Cruz, the IMF, and the Thurgau Institute of Economics. For helping me revise this monograph on short notice I wish to thank Veronica Schmiedeskamp, David Kaun, William Koch, and Andrew Verner.

Words do not suffice to describe the support of my wife Juliane. During all stages of the dissertation she gave me invaluable emotional backing and helpful advice to steer me through the adversities of a young academic career. It was she who always stood by my side and who willingly tolerated never ending working hours.

St. Gallen, January 2006

List of Symbols

B	Value of money market account
C	Coupon payment
c	Annual CDS premium
D	Macauley duration
d	Stochastic discount factor
F	Forward rate
g	Premium leg
h	Protection leg
L	Average life
N	Notional value
P	Risky coupon bond price
Q	Cumulative distribution function of the default probability
q	Unconditional default function
R	Recovery adjusted discount rate
r	Instantaneous risk-free rate
S	Survival function
s	Continuous risk spread
Y	Annual yield
y	Continuous yield
Z	Risky zero bond price
Λ	Risk-neutral binary hazard rate process
λ	Hazard function
φ	Total recovery value
ψ	Recovery fraction of market value (RMV)
ω	Recovery fraction of face value (RFV)
τ	Credit event time

Contents

1   **Introduction** 1
   1.1   Relevance of the Topic 3
      1.1.1   Relevance for Global Investment 3
      1.1.2   Relevance for Emerging Countries 5
      1.1.3   Relevance in Academic Research 8
   1.2   Research Subject and Methodology 9
      1.2.1   Subject 9
      1.2.2   Methodology 12
      1.2.3   Structure 13

2   **Sovereign Lending and Default** 15
   2.1   The pre-1990 Episode of Sovereign Lending 17
      2.1.1   Infancy and the "Golden Age" 17
      2.1.2   Bank Loans and Restructuring 19
      2.1.3   The Brady Plan 21
   2.2   The post-1990 Episode of Sovereign Lending 22
      2.2.1   Mexico 1994–1995 28
      2.2.2   The Asian Crisis 1996–1997 29
      2.2.3   Russia 1997–1998 30
      2.2.4   Brazil 1998–1999, 2002–2003 31
      2.2.5   Pakistan 1998–1999 32
      2.2.6   Ecuador 1998–1999 33
      2.2.7   Ukraine 1998–2000 34
      2.2.8   Turkey 2000–2001 35
      2.2.9   Argentina 2000–2005 36
      2.2.10 Uruguay 2001-2003 38
      2.2.11 Moldova 2002 39
      2.2.12 The Caribbean Restructurings 2005–2006 40

        2.2.13 Outlook 41
    2.3 The Theory of Sovereign Lending and Default 42
        2.3.1 The Theory of Lending 43
        2.3.2 Crisis Literature 45
        2.3.3 The Literature on the IMF's Role 47
    2.4 Empirical Evidence 49
        2.4.1 Determinants of Crises 50
        2.4.2 The Effect of IMF Involvement 52
        2.4.3 Determinants of Ratings 54
        2.4.4 Determinants of Spreads 55
    2.5 Concluding Remarks 58

3 Sovereign Restructuring 61
    3.1 Literature Review 62
        3.1.1 Sovereign Bankruptcy Procedures 62
        3.1.2 Analyzing Past Workouts 68
    3.2 Crisis Resolution in a Nutshell 69
        3.2.1 Liquidity and Solvency Crises 69
        3.2.2 Debt Swap, Soft and Hard Restructurings 71
    3.3 Evidence From Recent Restructurings 78
        3.3.1 Features of Recent Restructurings 80
        3.3.2 Resulting Present Value 88
    3.4 Lessons for Investors 96
        3.4.1 Investor Returns 96
        3.4.2 Modeling the Recovery Value 101
    3.5 Concluding Remarks 105

4 Modeling Sovereign Default Risk 109
    4.1 Literature Review 110
        4.1.1 Structural Models 111
        4.1.2 Reduced Form Models 112
        4.1.3 Recovery Schemes 116
        4.1.4 Outlook 118
    4.2 An Overture to Bond Analysis 119
        4.2.1 The Money Market Account and the Discount
              Factor 119
        4.2.2 The Price of a Risky Zero Bond 121
        4.2.3 The Price of a Risky Coupon Bond 122
        4.2.4 Yields, Spot and Forward Rates 124
        4.2.5 Default Probability Functions 126
        4.2.6 Bootstrap Analysis 127
        4.2.7 Bond Duration and Average Life 129

4.3 Functional Forms of the Term Structure.............. 131
    4.3.1 Affine Models 132
    4.3.2 Parsimonious Models.......................... 133
    4.3.3 Discussion 141
4.4 Modeling Recovery 143
    4.4.1 Recovery of Market Value 144
    4.4.2 Mixed Recovery 145
    4.4.3 Discussion 146
4.5 Empirical Implementation 147
4.6 Concluding Remarks............................... 150

5 **Empirical Estimations**............................. 153
5.1 Empirical Model Comparison 154
    5.1.1 The Nelson-Siegel Model 154
    5.1.2 Two-factor Nelson-Siegel With RFV 159
    5.1.3 The Weibull Model With RFV 163
    5.1.4 The Gumbel Model With RFV 168
    5.1.5 The Lognormal Model With RFV 171
    5.1.6 Discussion 176
5.2 Results From Other Countries 177
    5.2.1 Argentina 178
    5.2.2 Colombia 183
    5.2.3 Mexico 186
    5.2.4 Turkey 191
    5.2.5 Venezuela 194
5.3 Concluding Remarks............................... 197

6 **Credit Default Swaps** 199
6.1 An Introduction to CDS 199
    6.1.1 CDS Valuation 202
    6.1.2 The CDS Basis............................... 204
    6.1.3 The Role of Recovery 206
6.2 Empirical Evidence From Brazil 2002–2003 211
    6.2.1 Preliminary Data Analysis 211
    6.2.2 No Arbitrage With Two Instruments 214
    6.2.3 No Arbitrage With Three Instruments.......... 217
6.3 Concluding Remarks............................... 220

7 **Conclusion** 223

References 229

1

Introduction

What will the world look like thirty years from now? Is the US dollar going to retain its position as the key currency? Will the pace of global integration with growing trade and capital flows continue, or recede in a continuation of historical cycles? How will the industrialized world interact with countries emerging from underdevelopment and competing for scarce economic resources? What might be the impact of disruptive events like natural disasters or civil unrest?

Given such commonplace uncertainties, it might seem astonishing that one could plan for the future. Capital markets, however, seem to do so as investors are willing to buy bonds from all parts of the world promising payments up to forty years into the future. Despite the recent push for long-term investment opportunities, should not caution be recommended, in emerging markets and elsewhere? Is it rationally justifiable to invest in emerging country governments, many of which have a shaky past but are hardly ever called to account? Do investors lend their money in the firm belief that the debtor will obey its repayment obligations?

Such qualms, though, might not be warranted. When things start to turn sour, investors who missed a timely exit quickly become engaged in an effort to recover some lump sum. Just as a country is unlikely to disappear from the world map, it is practically impossible for a sovereign to shake off all past promises. However, with the large number of actors in this game and the different interests at play, it might appear miraculous how markets assign a value to sovereign risk.

The credit risk literature has tackled the issue of pricing default risk by means of sophisticated calculus based on expectations theory calibrated with empirical data. By doing so, modeling the term structure of risk spreads became the focal point. Postulating that the loss given default is a fraction of the market value helped to simplify matters greatly as the spread alone embodied all facets of credit risk. This dogma remained intact even for analyzing sovereign default risk where governments, in contrast to corporations, are immune to, and not liable for, bankruptcy proceedings.

On the one hand, today's scientific credit risk analysis has advanced to a level of abstraction beyond the grasp of many investors. On the other hand, market practitioners try to explain market movements with bits of new information arriving, often lacking consistency. However, both ways of understanding bond markets focus on bond yields and the term structure of risk spreads. This widespread practice is also reflected in credit risk management tools and the Basel II capital adequacy framework. The literature to date has not considered repricing of recovery value assumptions as an important feature of any market. The seminal paper by Duffie and Singleton (1999) helped to shape this attitude by advocating the recovery of market value concept. While handy and sufficient in most cases, this concept allows little flexibility in modeling recovery. Apparently, this accepted custom may soon change. Pan and Singleton (2005) acknowledge:

> "Equally central to modeling the credit risk of sovereign issuers
> is the recovery in the event of default".[1]

The starting point of this study addresses some of the shortcomings of conventional credit risk frameworks. This is done by adapting traditional models to better suit the valuation of bonds and credit default swaps (CDS) subject to sovereign risk. The short history of sovereign bond markets and the development of ad hoc approaches to sovereign debt crises mark this study as an early contribution in exploring the peculiarities of sovereign risk valuation. This monograph is intended for investors searching for a toolkit when investing in emerging market sovereign instruments, scholars interested in alternative ways of evaluating default probabilities and recovery values, as well as other curious readers intending to fend off urban legends.

The rest of the introduction explains why this topic has grown so relevant in recent times. It motivates the subject from three different perspectives, from both the buy and sell sides, as well as from an academic point of view. Section 1.2 of this chapter defines the scope of the study, provides explanatory commentary on the nature of empirical studies, and explains the structure of the following chapters.

---

[1] Pan and Singleton (2005), p. 1.

1.1 Relevance of the Topic

Along with the increase in global liquidity and market integration during the last decade, new investment opportunities arose around the world.[2] Investors strove for higher returns and diversified risk, creating demand and liquidity for new financial products. Developing countries matched this demand, exploiting different channels to access much needed international capital. Regulatory bodies, such as the International Monetary Fund (IMF), and academia devoted much effort to overseeing, steering, and analyzing this development. These three pillars are used in the following to highlight the relevance of sovereign bonds for emerging countries today.

1.1.1 Relevance for Global Investment

Globalization includes the boundless flow of capital around the globe. Citibank was among the pioneers of truly global banking, trying to break the ties of the U.S. banking regulation and investing in developing countries. Soon, other banks joined this development.[3] In the 1970s, a lending boom to emerging countries developed. Long-term syndicated loans to these countries became an important vehicle for recycling petrodollars.[4] Bond markets also internationalized, creating all sorts of internationally placed issues such as eurobonds, Yankee bonds, Samurai issues, and global bonds.[5] After the establishment of a market for trading high yield instruments of corporations during the 1980s, risky bond instruments found their way into the portfolios of institutional investors. In 1989, the Brady plan, which effectively securitized third world debt, paved the way for sovereign issuers into the high yield market. Today, external emerging market debt accounts for close to half of the high yield market (which has an estimated debt

---

[2] See Bekaert and Harvey (2003) and Erb et al. (1999).

[3] See Friedmann (1977).

[4] The pseudonym "petrodollar" denotes capital received by oil exporting countries after the oil price shocks. These funds were invested mainly in U.S. Treasury bills or transferred to American and Western European banks. Banks used these surpluses to increase their lending, especially to less developed countries, mainly in Latin America.

[5] Eurobonds are bonds issued on the euromarket, i.e. the international capital markets, in a currency other than that of the country of issuance. Yankee and Samurai issues are denominated in US dollar or yen, respectively, and are exclusively sold in the U.S. or Japanese markets. Global bonds are issued on the international markets like eurobonds, but may be denominated in the same currency as the country of issuance. International bonds serve as an umbrella term.

volume of more than one trillion dollars). The share of sovereign issues of all emerging market bonds is about two-thirds.[6]

Despite this expansion, global capital flows proved volatile during the last 15 years. Financing the needs of emerging countries was put into the hands of the international investor community while the recipient countries initially lacked the economic and institutional structures necessary to effectively manage large capital inflows and outflows. Allocating capital globally also meant an exposure to changes in global liquidity, fluctuations in global risk aversion, and contagion. Cross-over investors welcomed emerging market debt as a substitute when domestic returns in industrialized countries were slack. The resulting volatility is reflected in Table 1.1, which shows a summary of annual capital flows to thirty emerging countries. Whereas net capital flows from official lenders turned negative due to loan repayments, net private capital flows grew strongly, but remained volatile.

More recently, however, the sovereign bond market underwent a process referred to as "secular maturation". While fundamentals improved, the growing number of bonds outstanding established a relatively liquid asset class. The simultaneous emergence of a credit derivatives market substantially increased the smooth functioning of the bond market. This maturation enabled a widening of the investor base. Initially, only dedicated investors dealt in the market, exploiting their specific knowledge of emerging market bonds. In recent years, however, improving fundamentals and declining spreads have attracted a broader investor base. Large pension funds and buy-and-hold investors still make up the majority of investors while investors from other mandates (e.g., global bond funds) are being drawn into emerging market bonds on an "off-index" bet. Hedge funds make up about one-third of investments in emerging market debt and have become the marginal price setters.[7]

The maturation of the market affects not only the investor base, but the composition of emerging market debt as well. While external bond flows surpassed commercial bank loans by the end of the 1990s, this expansion is dwarfed by the growth of the domestic debt markets. Since 1996, the stock of domestic debt exceeds the amount of external debt in emerging markets and continues to increase at double digit rates. This trend might overcome the "original sin" problem of investors un-

---

[6] These numbers are compiled from Merrill Lynch, "Size and Structure of the World Bond Market: 2002", as well as JP Morgan Chase, "Emerging Markets Debt and Fiscal Indicators", July 2005, and "Emerging Markets Bond Index Monitor", June 2005.

[7] See JP Morgan Chase, "Emerging Markets As An Asset Class", October 2005.

Table 1.1. Capital flows to emerging market countries 1990–2004

Capital flows (US$ billion)	1990	1992	1994	1996	1998	2000	2002	2004†
Net external financing	**75**	**159**	**201**	**338**	**191**	**184**	**117**	**250**
Net private flows	*35*	*121*	*175*	*334*	*139*	*187*	*120*	*303*
Net equity flows	17	47	99	127	133	152	119	177
Net direct investment	14	31	65	92	121	139	112	138
Net portfolio investment	3	16	34	35	12	13	1	39
Net private credit flows	17	74	76	207	6	35	2	127
Net commercial banks	9	29	43	123	-55	-1	-4	54
Net nonbanks (mostly bonds)	8	45	33	84	61	36	5	73
Net official flows	*40*	*38*	*26*	*4*	*52*	*-4*	*-3*	*-28*
IFI	10	9	5	7	38	3	8	-19
Bilateral creditors	30	29	22	-3	14	-7	-11	-9

Source: Institute of International Finance. Figures for thirty emerging market countries representing more than 90% of net private flows to developing countries. The official sector consists of the international financial institutions (IFI), i.e. the IMF, the World Bank, and multilateral development banks, as well as bilateral lenders, i.e. other governments.
(†) Estimate.

willing to buy domestic currency instruments, and at the same time reduce the danger of external imbalances. Moreover, investors are increasingly looking into new instruments offering exposure to sovereign risk. These include asset securitization and structured instruments, as well as derivatives such as sovereign credit default swaps.[8] Given market incompleteness and a general lack of financial data on sovereigns, the pricing of such new instruments requires a profound knowledge of the nature of sovereign default risk, sovereign restructurings, and recovery values.

1.1.2 Relevance for Emerging Countries

Additional to these "push" factors, fundamentals in emerging countries have been growing stronger and present a clear "pull" effect. The recent contraction of sovereign spreads and the parallel surge in capital inflows are seen as a result of both effects. Figure 1.1 illustrates the course of

---

[8] See Ketkar and Ratha (2001), Alles (2001), and Packer and Suthiphongchai (2003).

sovereign spreads as represented in the JP Morgan Chase Emerging Market Bond Index Global (EMBI Global).[9]

Fig. 1.1. Emerging market bond spreads, 1990–2005

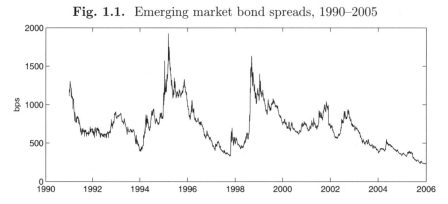

Source: JP Morgan Chase. Spliced history of JP Morgan EMBI and EMBI Global spread indices as provided by Bloomberg.

Learning from the crises of the last decade, emerging market governments now strive to improve their economies' external balances. When pooling all developing countries, the total current account balance turned positive in 2000 for the first time in decades. More countries have abandoned inflexible exchange rate arrangements—although some prominent examples did so only painfully—which had contributed to external imbalances in many cases. In recent years, more countries were able to accumulate a comfortable cushion of foreign reserves, extend the duration of their debt, and establish a smooth pattern of future installments.

However, as evident from Fig. 1.2, the share of sub-investment grade issuers within the emerging market asset class increased. The gradual improvement of credit ratings in some countries (in particular transition economies) failed to compensate for the large number of new entries with sub-investment grade ratings. The few success stories, such as the investment grade ratings of Mexico and Russia, were overshadowed by the number of debt crises and defaults, headed by Argentina's historical default in 2001.

---

[9] The EMBI Global is a total return index for US dollar denominated debt instruments issued by sovereigns and quasi-sovereign entities of emerging countries. As of July 2005, the index included 184 instruments such as Brady bonds, loans, and global bonds from 32 countries with a total market capitalization of $281 billion.

Fig. 1.2. Evolution of sovereign ratings in the EMBIG 1993–2005

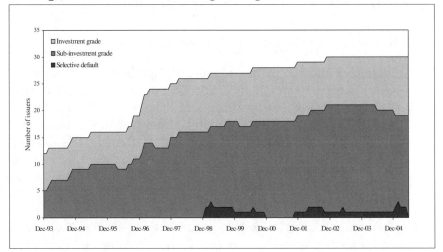

Source: derived from Standard & Poor's and JP Morgan Chase. Standard & Poor's long-term foreign currency rating of sovereign issuers represented in the EMBIG in 2005.

As long as markets maintain their recently acquired resilience, governments will continue to exploit the full potential of sovereign bond markets. Expertise from debt swaps is used to pre-finance bond issues with short remaining maturities and eliminate old Brady-style debt. Accredited debtors utilize such exchanges to concentrate liquidity on fewer issues, and introduce clauses in their debt contract which strengthen the issuer's rights and ensure a more orderly workout process. The concept of "private sector involvement" (PSI), promoting equal burden sharing among all creditors in debt crises, and the G-8 stance towards debt forgiveness provide a stage for future bond restructurings. Judicial support for the controversial write-off in the Argentine restructuring has fueled this development. A sophisticated knowledge of past transactions presents good ammunition in face of the innovations the financial market is likely to see, such as collective action rulings or repeated restructurings. After the wave of soft restructurings in the Caribbean region in 2005, similar deals are expected to become the prevalent credit event for sovereign debt instruments. The exceptionally high recovery values in such transactions present a challenge to traditional valuation models for default contingent claims, such as bonds and credit default swaps. A large portion of this study is dedicated to the issue of recovery specification, offering insight into the consequences of financial crises and sovereign restructurings for the financial markets.

1.1.3 Relevance in Academic Research

The valuation of risky debt instruments and their derivatives has received considerable attention in academic research. The most basic theoretical foundations laid by Black and Scholes (1972) and Merton (1974) have been extended to a comprehensive framework of theoretical and empirical elaborations on credit risk.[10] Driving this development is the need for more sophisticated risk management, along with the expansion of the credit derivative market, and the advancement of structured products such as Collateralized Debt Obligations (CDO). Thanks to the Basel II Capital Accord, recovery risk now enjoys closer academic scrutiny.[11] At the same time, more theoretical and empirical work is dedicated to emerging market finance.[12] After close attention to sovereign debt contracts during the debt crises of the 1980s, a recent strand of the literature has been devoted to analyzing sovereign credit risk in traded instruments, foremost bonds. Chapter 2 provides a closer look at this expanding field.

The focal point of the empirical literature in this field, however, is the term structure analysis and its relation to the fundamentals. Given the existence of sufficient time series data on different bonds subject to the same credit risk, traded instruments issued by a sovereign present a well suited research object for this purpose. The financial literature has exploited these data by means of sophisticated stochastic term structure models, such as Duffie et al. (2003).

However, stochastic models are currently not suited to incorporate different concepts of recovery, thus foregoing an important aspect of sovereign restructurings. Empirical models, although often lacking the sound theoretical foundation of arbitrage free modeling, offer greater flexibility while presenting no lesser a fit to the data. By compromising on the no-arbitrage argument, the empirical part of this study yields weekly estimates of the recovery value implied in sovereign bonds and credit default swaps. This is intended to close a gap between the economic literature on fundamental determinants of crises and the asset pricing literature. The analysis assesses recent sovereign restructurings

---

[10] See, among others, Ammann (1999), Cossin and Pirotte (2001), and Bielecki and Rutkowski (2002).

[11] See, as a collection of recent articles, Altman et al. (2005).

[12] See Bekaert and Harvey (2002, 2003).

through the lens of investors and implements the findings in a risk-neutral bond valuation model which differs from the current mantra of credit risk modeling. The empirical part yields estimates of unobservable variables which, in contrast to previous contributions, separate default intensity parameters from the estimated recovery rate. Such measures might prove useful for future economic research on fundamental determinants of sovereign solvency and liquidity which, up to now, frequently relies on mere simple indices (such as EMBIG spreads) of sovereign risk.

This study also differs from the existing literature by concentrating on countries and periods of financial distress. When default is close, the price of a bond is dominated by the legal rank of the contractual claim rather than the expected value of future cash flows. In the traders' lingo, bonds are referred to as "trading on a price basis" instead of a "yield basis", as traditional term structure models assume. By addressing this appropriately, a better fit of empirically observed bond prices is obtained. This evidence diverges from the irrelevance presumption of recovery implicit in the recovery of market value (RMV) assumption.[13] Chapter 6 elaborates on this by illustrating the relevance of the recovery value when pricing credit default swaps during financial distress.

These points comprise the main contribution to the existing literature. The results will prove useful for scholars of sovereign risk and sovereign bond investors alike. While the empirical results are intended to give an idea of how sovereign risk analysis can be approached, the conceptual issues addressed will help to conduct a more sound analysis of sovereign investment instruments in the future.

1.2 Research Subject and Methodology

To provide the reader with some guidance on what to expect from reading this study, the following explains which topics are addressed in this study, which methodology is applied, and how the remainder is structured.

1.2.1 Subject

Sovereign risk refers to the possibility that a sovereign government (or its responsible entity) fails to fulfill a contractual obligation such as a

---

[13] See Duffie and Singleton (1999).

debt contract. Any kind of breach or change of the contractual clauses are referred to as default. Such a broad definition of default does not necessarily correspond to economic insolvency or judicial bankruptcy known from corporations. Since a sovereign cannot go bankrupt in the traditional sense, the occurrence of default mostly goes back to some political decision. When the term default is used within this study, however, it is intended to carry a neutral meaning in the sense of a "credit event". While it is true that default can lead to write-offs, a debt restructuring may in fact benefit debtors and creditors alike. Chapter 3 elaborates on this.

Unique to this examination is that the default risk is borne by a sovereign. This distinguishes this study from the large body of literature on corporate default risk. However, the concepts of modeling sovereign versus corporate credit risk are related. Although there is no bankruptcy court for sovereigns, the scope of sovereign immunity has always been limited, given the financial interests at play. In the past, such limitations were enforced by means of gunboat diplomacy or trade restrictions. Today, such measures have been replaced by applied law on the international financial markets.[14] What remains unique is the sovereign's de facto leeway in renegotiating the debt. This process is ruled by political realities, rather formal legal principles. In comparison to corporate defaults (which occur in larger numbers and provide a comparatively homogeneous data set on historical default and recovery rates), this attribute presents a challenge to any empirical assessment.

The study of international sovereign bonds sheds light on the main aspect of sovereign default risk. Creditworthiness on the international financial markets is regarded as a crucial condition for participation in the global economy. The importance of international capital flows for growth and development is an acknowledged fact in both theoretical and empirical economics. The credit standing of a country's government thereby provides some limits on the creditworthiness of other economic entities located in that country. Emerging market governments therefore have a strong incentive to demonstrate their qualities as debtors and compete for international capital. This contrasts sovereign debt issued in domestic capital markets (i.e. within those countries and under domestic legislation), which are typically less mature and vary in the degree of their rule of law. Compared to domestic markets, this study

---

[14] Certainly, some caveats apply. While it is possible to reach a judicial ruling against a sovereign in specific cases, the main challenge to the plaintiff consists in the enforcement of such a ruling. Andritzky (2004b) contains a short review of the legal issues at work and provides further references.

of the international bond markets enjoys the virtue of homogeneity in the research subject. Global bonds, i.e. bonds sold in an international investor universe under the law of an acknowledged foreign financial center, are issued under fairly comparable contractual terms and are traded among a comparatively homogenous community of investors. Since sovereign debtors often borrow large sums, their international bonded debt is often split into several issues. While there are differences in terms of maturity, coupons, and amortization, these claims share the same seniority, i.e. rank pari passu in case of a credit event. Market prices of global bonds therefore contain rich information about sovereign risk and facilitate the comparison between countries. This focus allows for an analysis of the following questions:

1. What were the intentions and the result of recent crisis resolution efforts which involved sovereign debt problems, given the idiosyncratic nature of their circumstances?
2. What are the effects of sovereign restructurings from an investor's perspective, both in terms of returns on investment and implications on modeling recovery?
3. How can sovereign default risk be modeled, and what information can be extracted from the market prices of bonds?
4. What is the performance of such a model in comparison to traditional models when applied to empirical data?
5. To what extent does the recovery assumption matter for pricing credit default swaps, especially during distress?
6. What information about the expected recovery can be revealed from market quotes of credit default swaps?

Concentrating on these questions requires a strict demarcation of several considerations that cannot be addressed in this study. The focus of this study suggests limiting the empirical analysis to debtors who embody considerable risk of default. While this does not question the general applicability of the model developed here, insightful empirical results can only be obtained when bond prices are subject to a substantial threat of default. A second limitation is dictated by the availability of data. While some countries (independent of their size) have a large number of bonds outstanding, others might use different channels of financing and therefore do not find their way into this analysis. Furthermore, the quality of the financial data is not always sufficient, as trading volumes are thin or financial time series were not obtainable for this study.

The cumulative effect of these aspects motivates the country selection for the empirical parts in Chaps. 5 and 6. The analysis is conducted for fixed coupon bonds denominated in US dollars. This is the dominant currency of denomination, and most issuers (except for some Eastern European states) denote the largest portion of sovereign bonds in dollars. However, given a sufficient sample of euro, sterling, or yen denominated issues, the proposed model can be applied in a similar fashion. A mixed sample of bonds of different denomination requires disentangling currency risk from bond spreads though—an exercise beyond the scope of this study. Floating rate notes make up an insignificant portion of the market, but could be modeled in a similar manner. Due to their declining importance, this study also foregoes the analysis of Brady bonds.[15] Semi-sovereign issues, such as bonds from public companies or bonds with sovereign guarantees, may also show significantly different characteristics, and do not belong in this study. Only with these very strict limitations was it possible to achieve a homogeneous sample of sovereign bonds. It also serves to reduce potential distortions originated by trading illiquidity or market segmentation, topics which will appear as side aspects only.

1.2.2 Methodology

For most of this study, the research approach is straightforward and laid out in the introduction to each chapter. Solely for Chaps. 5 and 6, which contain empirical estimations, the following thoughts highlight the benefits and limitations of such an endeavor.

The theory of asset pricing provides the foundation to derive the present value of future payments under the veil of default risk. Even after considering fundamental rules of stochastic calculus and no arbitrage, there remain several ways to value claims subject to default risk.

While this will be discussed in depth in Chap. 4, the following focuses on the implications of interpreting the empirical results. Empirical estimates are known to be a test of both the respective hypothesis, as well as the model applied with all its underlying assumptions. Sound theory can suggest a pricing model, arguing that this is the true way to determine fair asset prices. This is called the normative view. If consensus exists on such a unique model, it is justified to determine the endogenous variables from asset prices. The outcome reflects the positive view of things. As soon as there are differing views on what the "true model" should look like, however, the result of an empirical calibration might become the product of its assumptions.

---

[15] On Brady bonds, see instead Bhanot (1998), Buckley (2004), Claessens and Pennacchi (1996), Izvorski (1998), Pages (2001), and others.

Reduced form models of credit risk, as applied in the empirical part of the study, are particularly vulnerable to these concerns. The decisive parameters of credit risk frameworks such as default probability, recovery expectation, and risk aversion, present unobservable parameters which can only be proxied by some measurable variable. As long as market participants do not share a common view on either the model parameters or on the model itself, market prices might give only a blurred picture of what theoretically is considered as credit risk.

A proof of plausibility for empirically estimated measures of credit risk is therefore advisable. The sound theoretical foundation of the pricing and estimation model, together with a check of the underlying assumptions, is a useful starting point. Furthermore, some guidance can be derived from historical experience, even if past events cannot directly be compared to expected future events. Chapters 2 and 3 provide this background. Another approach is to determine the empirical fit of the model with the data, an aspect highlighted in Chap. 5.

This caveat of empirical research is stressed in the face of missing benchmarks of credit risk parameters. The term structure curve of credit risk spreads, for example, is an abstract measure contingent upon a set of assumptions. Making justified manipulations to these assumptions might yield a different, perhaps unfamiliar curve which can nevertheless be similarly plausible. This has to be kept in mind for the empirical estimations in this study. The plausibility of a different view on modeling the loss given default is a renunciation of current conventions. These considerations, however, are warranted by economic intuition, historical experience, and market practice.

1.2.3 Structure

This monograph has seven chapters. After this introduction, Chap. 2 reviews the history of sovereign lending and default. The first part focuses on bond lending a century ago, showing how sovereign default and bond restructurings were handled then. Historical evidence provides early examples of bondholder coordination and collateralization, features which are discussed again today. The second part reviews the development which led to the new age of bond lending in the 1990s, and offers a short outline of recent financial crises in emerging markets. Furthermore, Chap. 2 reviews the literature on sovereign lending, financial crises, and the international financial institutions. The section presents current theories, positions this study within the literature, and offers further reference.

Chapter 3 analyzes recent restructurings of sovereign debt—such as the Argentine mega-restructuring in 2005—from the point of view of an investor. The empirical evidence supports the distinction between "soft" and "hard" restructurings, depending on how advantageous the restructuring deal is for bondholders. The heterogeneity of such a workout deal is reflected in the wide range of resulting recovery values.

The following part links this finding to existing bond pricing frameworks, considering two kinds of recovery assumptions. Chapters 4 and 5 are devoted to the analysis of global sovereign bonds issued by large emerging countries. In a first step, Chap. 4 gives an introduction to financial calculus and elaborates on ways of modeling sovereign default risk. For modeling recovery, a framework mixing the recovery of market value (RMV) and the recovery of face value (RFV) approach is suggested. Applying different variations of term structure models, Chap. 5 evaluates these approaches in a case study of Brazilian global bonds. The next section extends the analysis to half a dozen countries. The results provide a set of estimates on implied parameters, such as the risk-neutral default intensity and recovery rate.

Chapter 6 advances into derivative markets, showing the relevance of recovery assumptions on the pricing of credit default swaps. An analysis of the Brazil crisis 2002/2003 yields a differentiated picture of expected recovery values, showing why protection becomes so expensive when soft restructurings are the prevalent path to crisis resolution. A joint model of bond prices and credit default swap spreads is used to extract market implied recovery values, distinguishing RMV and RFV parameters.

The last chapter, Chap. 7, presents a synthesis of the results.

2

Sovereign Lending and Default

"Countries don't go bankrupt."

This is a famous saying in the financial community. Despite waves of sovereign defaults and restructurings, the statement is still true at its core. The reason for this is to be found in the concept of sovereignty, the prevalent principle of today's world order. Within sovereignty, two complementary dimensions are inherent, the internal and the external dimension.

The internal dimension of sovereignty is constituted by the supreme authority of a country's political body. The ruling institution—constitutional governments and parliaments, dictators and juntas, monarchs and the like—assumes the power of legislative actions, fiscal budgeting, and overall economic policymaking. This authority embraces the decision to reach out to international capital markets and likewise secures sufficient revenues for debt service. Thomas Hobbes drafted this notion of unlimited internal power, or his "Leviathan".[1] In theory, the Leviathan can exercise the power to raise sufficient means for external debt service: he might introduce capital controls, increase taxes, nationalize the corporate sector, or pledge the country's assets to foreign creditors. In very few areas the sovereign's discretion is limited by conventions of international law (for instance human rights, encompassing only basic economic privileges such as property rights), but weak enforcement renders the scope of protection even smaller. This is why countries can rarely go "bankrupt".

The external dimension of sovereignty provides protection against foreign influence. This idea of non-intervention in national affairs was commenced by the Peace of Augsburg in 1555 ("cuius regio, eius reli-

---

[1] See Hobbes (1968).

gio") and later the Peace of Westphalia in 1648. Today, Article Two of
the United Nations Charter enshrines the "political independence and
territorial integrity" and leaves few exceptions justifying interventions.[2]
This leaves few levers for bondholders of sovereign debt against repudiation. However, the lack of internationally agreed rules on sovereign
lending, default, and restructuring can be seen as logical complement to
the lack of effective protection of human economic rights. This aspect
notwithstanding, external sovereignty today is confined by the denial
of absolute sovereign immunity. Diplomatic immunity, as manifested in
the Vienna Convention of 1961, excludes sovereign acts from any foreign legislation ("ius imperii"). International debt, even when issued by
a sovereign entity, is subsumed as business activity ("ius gestionis"),
enjoying limited immunity. National law, such as the U.S. "Foreign
Sovereign Immunities Act" of 1976 or the British "State Immunity
Act" of 1978, has fleshed out this notion.[3] However, little legal certainty has made litigation against sovereign debtors a costly endeavor
with few chances for success.[4]

Although the decision to default and restructure sovereign debt is,
seemingly, at the issuer's discretion, political pressure from the outside
plays a significant role. Whether it is the U.S. Treasury pushing for a
bail-out of Mexico in 1994, the IMF agreeing on Turkey's aid package,
or Venezuela's Hugo Chávez coming to the aid of his fellow South American leaders by buying their bonds, international politics are always at
the forefront of crises. Sovereign bond investors and their interests play
a subordinate role in this play, at least when lacking a sound com-
mand of lobbying power. The threat of litigation has hardly impressed
any debtor government. Domestic sovereign and sub-sovereign bonds, in
contrast, are very distinctive in this regard.[5] They underly primarily the
domestic legislature (whose idiosyncratic nature hampers cross-country
comparisons), although domestic markets are increasingly interlinked
with international markets. While holders of international sovereign

---

[2] However, such doubtable interventions have frequently occurred since the end of
the Cold War, although never in response to sovereign debt repudiation. Indeed,
quite the reverse causality prevailed when Pakistan faced economic sanctions in
response to atomic tests, forcing the country into a restructuring of sovereign
bonds in 1999.

[3] Waivers of immunity to foreign creditors, common in sovereign bond contracts,
enforce only this aspect and, as of today's legal doctrine and judical practice, do
not undermine diplomatic immunity.

[4] See Andritzky (2004b). However, returns on litigation, when successful, can be
substantial. See Singh (2003b).

[5] On the restructuring of subnational debt see Schwarcz (2004).

bonds are off the reach of internal sovereign power, holders of domestic bonds are not. Their claims can theoretically be repudiated by the legislature in the wink of an eye.

This chapter is dedicated to exploring the past and current states of sovereign debt by looking at its history and reviewing the theoretical and empirical literature. Knowledge of past lending arrangements and recent crises helps to understand the manifoldness of this topic, especially since the patterns of lending and crises evolved over time. The theoretical scientific literature provides common frameworks which strive to reproduce the main characteristics of sovereign lending. Some underpinnings are provided by the empirical literature, although methodological problems and the idiosyncratic nature of events pose serious caveats. While the review in Sects. 2.3 and 2.4 focuses on the economic literature, more specific references to the asset pricing and credit risk literature follow in Chap. 4. The literature specific to sovereign restructurings and credit default swaps is discussed in the respective chapters (see Chaps. 3 and 6).

2.1 The pre-1990 Episode of Sovereign Lending

Although today's modern global sovereign lending is somewhat unique, it would be negligent to assume that there is nothing to learn from the past. Foreign lending (in the form of both bonds and loans), sovereign default, and sovereign restructurings occurred before, even in the 19th century (see Fig. 2.1). Wave-like patterns of international lending and default were present then and now. The following presents a short wrap-up of the events of that time.

2.1.1 Infancy and the "Golden Age"

The first well documented wave of lending in modern times occurred in the post-Napoleonic era of the 1820s with flows especially directed to newly independent countries in Latin America. A wave of defaults followed soon. Another high point was reached in the 1850s with subsequent defaults concentrated in Latin America and the Mediterranean. The period between 1870 and World War I marked the truly "golden age" of sovereign lending, with significant international capital flows into emerging markets and a degree of financial integration only regained again a century later.[6]

---

[6] See Lindert and Morton (1989), Sachs and Warner (1995), Aggarwal (1996), Bordo et al. (1998), and O'Rourke and Williamson (1998).

Fig. 2.1. Historic sovereign default rates 1820–2000

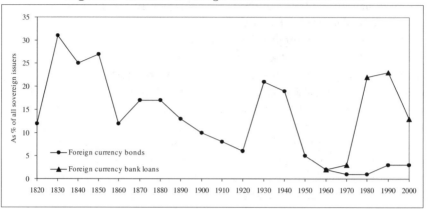

Source: Standard & Poor's.

In this early era of foreign lending, foreign currency bonds were the main lending vehicle while bank lending was limited to short-term trade finance and interbank credit lines. Trading took mostly place in the London bond market with traded government loans denominated in pounds sterling.[7] Liquidity was astonishingly high with the total trade value accounting for about half of Britain's GDP, a mark unmatched until today.[8] Certainly, the "emerging markets" of that era were different ones from today's. While some of today's emerging countries did not even exist on the world map, first world nations like Japan once belonged to that class (e.g., in terms of relative per capita income or capital inflows). When comparing this episode with the post-1990 era of emerging market bond lending, Mauro et al. (2002) find that the former period suffered from both fewer country crises, as well as few global jitters. Spreads were considerably less volatile and, notwithstanding economic fundamentals, generally lower than today.

The situation then differed much from today in two more respects.[9] First, sovereign immunity constituted a much stronger stance as there was no distinction between "ius gestionis" and "ius imperii" at that time. Second, debt instruments mostly came with identifiable collaterals (such as railroads) or were backed by the earmarking of tax rev-

---

[7] See Fishlow (1985), p. 394.

[8] The Emerging Market Traders Association estimates the trading volume of emerging market debt to about $4 trillion in 2003, which—compared to the economies of the two major financial centers—corresponds to twice the GDP of the UK but only one-third of GDP for the U.S.

[9] See Mauro and Yafeh (2003).

enues. Today, collateralization is rare and occurs mostly in foreign loans to (public) corporations and, by a very limited volume, in project financing bonds.[10]

The workout of defaults at that time was facilitated by the Corporation of Foreign Bondholders (CFB), established to overcome coordination problems among British investors in 1868. This organization apparently achieved a very good track record, quickly reaching agreements on a number of default cases, most of them occurring in a wave during the 1880s and 1890s. The resolution periods of default events fell from an average of fourteen years (1821–1870) to six years (1871–1925). For those countries where a mutual agreement was not reached—mostly smaller ones, especially in Latin America and southern U.S. states— the common perspective was perpetual default.[11] However, Mauro and Yafeh (2003) argue that the CFB might not have been the central catalyst for workouts, since American bondholders (although lacking a similar institution) were far more successful in negotiating with those countries. Instead, the authors suspect that political affairs (e.g., the Monroe Doctrine) or trade links played a pivotal role in inducing an agreement on sovereign restructurings.[12]

A workout agreement was reached within the CFB through debt-equity swaps or the seizing of customs or tax revenues.[13] Given the existence of a collateral, debt-equity swaps resulted in the takeover of the underlying assets (such as land or railroads) or export goods. (Peru, for instance, offered guano.) Debt write downs, however, remained an exception for mutually agreed deals.[14]

2.1.2 Bank Loans and Restructuring

With World War I and the Great Depression, many formerly well-situated debtors suffered from repayment problems. This triggered a

---

[10] See Dailami and Hauswald (2003).

[11] On sovereign debt crises in U.S. states during the 1840s, see Sylla and Wallis (1998) and English (1996), as well as Weidenmier (2005) for Confederacy debt.

[12] However, Wright (2002) highlights another function of the bondholder committees, namely reducing information asymmetries.

[13] See Eichengreen and Portes (1989b), and Mauro and Yafeh (2003).

[14] There is some evidence of debt write downs, mostly under extraordinary circumstances. Lindert and Morton (1989) mention Mexico under Porfirio Diaz in 1885, and the Romero Plan in Argentina in 1893. The Mexican refunding loans of 1898 and 1914 could also be interpreted as rescheduling agreements. Some of the large political turmoils in the early 20th century, such as the Mexican and Russian revolution and the break-up of the Ottoman Empire, resulted in total losses to creditors.

broad wave of defaults, affecting, besides Latin America, most of Eastern Europe, Turkey, and China.[15] The sweeping events of that time did not leave much space for debt renegotiations; investors simply accepted the massive losses. World War II and the following Bretton Woods era basically closed this chapter, and commercial cross-boarder finance dried up for a while. Only a few countries worldwide (some Western European and Asian countries as well as many Commonwealth nations) did not encounter a sovereign default event within this early age of sovereign lending.

The Bretton Woods conference established multilateral lenders (the IMF, the International Bank for Reconstruction and Development, as well as regional development banks) as new players on the scene. During the period from 1944 to 1971, hard currency flows to less developed countries (LDC) were mostly originated by both multilateral institutions and bilateral lenders. In 1956, bilateral lenders banded together in the Paris Club. Ever since, the Paris Club played a central role for developing and codifying policies regarding debt restructuring and debt forgiveness. Its principles, procedures, and terms broke ground for orderly workouts. Over time, standard treatments for debt renegotiations in qualifying countries ran the gamut from rescheduling to generous debt cancellation under the Initiative for Highly Indebted Poor Countries (HIPC).[16]

Commercial lending, however, revived in the 1970s due to "petrodollar recycling".[17] The wave of lending soon entailed a backlash of defaults.[18] Over time, the Bank Advisory Committee (or "London Club") was formed, born out of the necessity to somehow coordinate the following restructurings of commercial loans. The conditions of such restructurings were much more flexible, and deals tended to become more and more complex.

The debtor side equally engaged in positioning their interests. The North-South Dialog in the 1970s, initiated by the newly formed Group of Seventy-Seven (G-77) of less-developed countries, led to a somewhat more benign approach to sovereign debt restructuring. After the

---

[15] See Eichengreen and Portes (1986, 1989*a*).

[16] For a more complete review, see e.g. Rieffel (2003).

[17] As a result of the oil price increases, oil exporters deposited their revenues at major international banks. With excess liquidity in markets like the U.S. and increasing demand for loans from oil-importing developing countries, a lending boom to these countries unfolded. Citibank was the leading institution, but others quickly jumped on the bandwagon.

[18] Lindert and Morton (1989) offer striking evidence that the same countries got caught in debt problems as during the previous heydays of commercial lending.

systemic crises which enfolded in Latin America in the 1980s, the resolution process first got stuck in a lingering process of rescheduling, but finally moved to gradually more generous debt forgiveness. In the end, traded instruments and the Brady plan opened a new vehicle for commercial lending (and handling workouts) in the late 1980s.

2.1.3 The Brady Plan

Table 2.1. Brady restructurings

Country	Date of exchange	Amount 1/ (US$ bln)	Recovery 2/ (percent)
Mexico 3/	May 1988	0.6	n/a
Mexico	March 1990	48.2	65
Costa Rica	May 1990	1.6	16
Venezuela	December 1990	20.6	70
Uruguay	December 1990	1.6	56
Nigeria	January 1992	5.3	40
Philippines	December 1992	5.7	50
Argentina	April 1993	28.6	65
Jordan	December 1993	0.9	65
Brazil	April 1994	48.0	65
Bulgaria	July 1994	8.3	38
Dominican Republic	August 1994	1.2	65
Poland	October 1994	14.4	48
Ecuador	February 1995	7.8	55
Panama	May 1996	3.9	55
Peru	November 1996	8.0	55
Côte d'Ivoire	May 1997	6.5	24
Vietnam	December 1997	0.8	60
Total		212.0	60

Sources: JP Morgan Chase, World Bank (2002), Rieffel (2003).
n/a: not applicable. 1/ Face value plus past-due interest. 2/ Approximate calculation by Rieffel (2003) based on figures provided in World Bank (2002). 3/ Aztec exchange creating the Aztec 2008 bond.

By the end of the 1980s, selected commercial banks started trading with emerging market loans. In 1988, the first loans-for-bond exchange was conducted by Mexico, creating the "Aztec" bonds with the principal value fully collateralized by U.S. Treasury bonds. Similar restructurings were announced by Brazil in September 1988.

These initial deals were soon succeeded by the launch of the Brady plan, named after the U.S. Treasury Secretary Nicholas F. Brady who negotiated the first formal Brady deal with Mexico early in 1990. This transaction was followed by a sweeping number of similar agreements all around the world (see Tab. 2.1). The central feature of the Brady plan consisted of the broadening of the investors' universe by securitizing the loans and issuing bonds on the international capital market. Other features included the reduction of the creditor's burden of debt service by increasing the maturity, and reducing principal and interest rate. Additionally, the formed securities included a collateralized principal and limited interest guarantees.[19]

The creation of Brady bonds quickened the investors' appetite for emerging market sovereign bonds. Issuance of foreign currency bonds traded on international capital markets picked up in the mid-1990s, enabling sovereigns with a lukewarm credit standing to tap into international bond markets. Governments quickly learned how to exploit this vast source for much needed foreign currency.

2.2 The post-1990 Episode of Sovereign Lending

Fig. 2.2. Issuance activity of international bonds by emerging market debtors

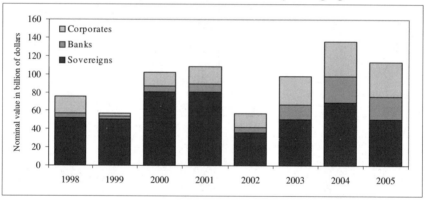

Source: JP Morgan Chase.

Progressive capital account liberalization and overall global financial integration supported the formation of a market for high-yielding sovereign bonds (see Fig. 2.2). For external financing, sovereign bonds broadly superseded commercial loans. Similarly, a trend from floating

---

[19] See Buckley (2004) for an analysis of these features.

towards fixed rate instruments commenced.[20] While issuing activity remains volatile (depending on global circumstances and rollover needs), the maturation of the market has allowed more and more non-sovereign lenders, such as banks and corporations, to exploit this source of external financing.

Fig. 2.3. Constituents of the EMBI Global

Country	Market value (US$ bln)	Spread (bps)	
Brazil	54.0	397	19.2%
Mexico	51.3	165	18.2%
Russia	37.2	138	13.2%
Turkey	21.4	268	7.6%
Venezuela	17.1	430	6.1%
Philippines	15.4	428	5.5%
Malaysia	9.5	78	3.4%
Colombia	9.5	316	3.4%
Peru	6.9	218	2.5%
Panama	5.7	220	2.0%
China	5.3	56	1.9%
Argentina	5.1	413	1.8%
Chile	4.7	56	1.7%
South Africa	4.6	96	1.6%
Lebanon	4.5	294	1.6%
Ecuador	3.5	735	1.2%
Ukraine	3.5	174	1.2%
Poland	3.3	45	1.2%
Uruguay	3.1	366	1.1%
Nigeria	3.0	348	1.0%
El Salvador	2.2	222	0.8%
Indonesia	2.0	265	0.7%
Bulgaria	1.6	71	0.6%
Hungary	1.5	46	0.5%
Egypt	1.2	61	0.4%
Serbia	0.9	277	0.3%
Morocco	0.8	134	0.3%
Tunisia	0.7	75	0.3%
Thailand	0.6	47	0.2%
Dominican Rep.	0.6	389	0.2%
Pakistan	0.5	173	0.2%
Côte d'Ivoire	0.2	3105	0.1%
Sum	281.4		

Source: JP Morgan Chase ("EMBI Monitor", July 2005).

A snapshot of the JP Morgan's EMBI Global constituents illustrates the dominance of Latin American issuers by today's market capitalization (see Fig. 2.3). While this ties in with history, more countries than ever are present on the market today. However, some sovereigns still lack market access, or rely on other sources of financing. The nominal total of all external sovereign bonds from emerging country issuers is estimated to have crossed the $400 billion mark.[21] This growth is out-

---

[20] See Borensztein et al. (2004), p. 10.
[21] See Merrill Lynch, "Size and Structure of the World Bond Market: 2002".

paced by the rise of domestic debt, reaching a total volume of about three times the external debt, and the surge of corporate bond issues. While this diminishes the importance of international sovereign bonds as a means of emerging market financing, it also reduces the "original sin" problem, a highly welcome development.

Clearly, the access to international debt markets in the 1990s benefited many emerging countries and can be seen as a clear precursor of the development of domestic debt markets. Unfortunately, debt problems did not stay away from the scene for long. Figure 2.4 gives an overview of the current financial architecture and its forums for resolving debt problems. IMF loans are only granted to member governments and enjoy the highest seniority, also reflecting the institution's function as lender of last resort. Loans of other multilateral banks (often granted in connection to IMF lending) are treated similarly. The moderate lending rates charged on multilateral loans reflect their seniority as emerging countries do not receive debt relief on these obligations.[22] The respective national bankruptcy code is applied to all private sector debt unless some kind of government guarantee is assumed, for instance for public sector debtors. While the treatment of bilateral and commercial debt mostly follows standardized procedures in the Paris and (with a lesser degree of formalization) the London Club, no such practice exists for bonded debt yet. The restructuring of external sovereign bonds thereby represent a sub-segment which attracted substantial interest from the political architects in recent times.[23]

Due to its connection to international financial markets and their volatile flows of capital, sovereign debt often stood at the center of post-1990 financial crises. While crises in the pre-1990 period mostly stemmed from current account imbalances, the new generation became known as capital account crises. However, a number of similarities apply. Both imbalances can lead to full-blown economic crises and affect not only the public, but also the banking and corporate sector. Table 2.2 gives an overview of the most prominent crises, some of which are described in the following subsections.[24] Besides learning about the

---

[22] On the debt repayment performance on IMF loans, see Aylward and Thorne (1998).

[23] See, for anecdotal evidence, the communiqué of the G-20 working group of major sovereign issuers and global leaders of private finance on the "Principles for Stable Capital Flows and Fair Debt Restructuring in Emerging Markets" from March 2005.

[24] For further reading and more references see Roubini and Setser (2004a), Dooley and Frankel (2003), Feldstein (2002), Edwards and Frankel (2002), and others. The descriptions below also draw on IMF country reports or IMF Public In-

Fig. 2.4. Restructuring forums for international debt

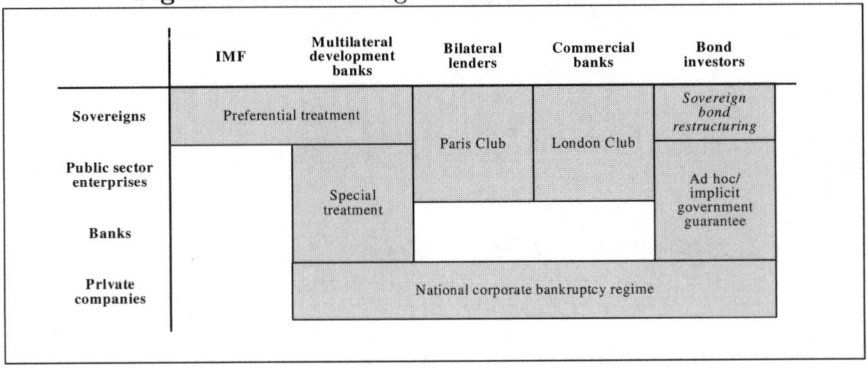

	IMF	Multilateral development banks	Bilateral lenders	Commercial banks	Bond investors
Sovereigns	Preferential treatment		Paris Club	London Club	*Sovereign bond restructuring*
Public sector enterprises		Special treatment			Ad hoc/ implicit government guarantee
Banks					
Private companies		National corporate bankruptcy regime			

Source: Rieffel (2003).

circumstances of each crisis, the review yields insights into the role of sovereign bonds and the effect of IMF intervention. Special attention is brought to the distinction of liquidity and solvency crises with regard to external public debt. In case of insolvency, the debt load is regarded as too large to be sustained in the long run, while illiquidity refers to situations where (even at moderate debt levels) a rollover of maturing debt is rendered impossible.[25]

---

formation Notes as well as other sources where indicated. Details of sovereign restructuring deals are discussed in depth in Chap. 3.

[25] See Sect. 3.2.1 for an advanced discussion.

Table 2.2. Overview of major post-1990 crises

	Mexico 1994-1995	Korea 1996-1997	Thailand 1996-1997	Indonesia 1996-1997	Russia 1997-1998	Brazil 1998-1999	Pakistan 1998-1999
Pre-crisis debt indicators							
External debt							
as percentage of GDP	33	31	60	43	35	31	68
as percentage of exports	196	104	150	164	140	369	347
Short-term foreign debt (percent of reserves)	203	289	136	158	255	126	189
Public debt							
as percentage of GDP	35	12	5	24	53	48	103
as percentage of revenues	155	58	27	186	148	143	646
Foreign-currency debt (percent of total)	53	n/a	12	100	60	n/a	52
Pre-crisis memorandum items							
Current account deficit (percent of GDP)	-7.1	-4.4	-8.1	-3.4	0.5	-4.3	-3.6
Fiscal deficit	-0.2	n/a	1.7	1.6	-7.6	-6.3	-6.2
Primary balance	2.1	0.6	2.6	2.9	-2.8	0	0.5
Interest payments (percent of revenues)	n/a	n/a	1.8	11	13	20	40
Exchange rate peg 1/	Yes	Yes	Yes	Yes	Yes	Yes	No
Crisis characteristics							
Currency crisis	Yes	Yes	Yes	Yes	Yes	Yes	No
Banking crisis	Yes	Yes	Yes	Yes	Yes	No	No
Corporate financial crisis	Yes	Yes	Yes	Yes	Yes	No	No
Output fall	Large	Large	Large	Very large	Small	Small	Modest
Crisis resolution							
IMF and bilateral commitment (US$ bln)	38.9	40.9	14.0	26.3	15.1 2/	32.9	1.1 3/
IMF and bilateral commitment (percent of GDP)	9.6	7.8	7.8	11.4	3.6	4.1	1.7
Sovereign debt workout							
Bond restructuring	No	No	No	No	Partly	No	Yes
Bilateral loan restructuring	No	No	No	Yes	Yes	No	Yes
Commercial loan restructuring	No	Yes	Partly	Partly	Partly	Yes	Yes

(continued on next page)

Table 2.2. Overview of major post-1990 crises (*continued*)

	Ecuador 1998-1999	Ukraine 1998-2000	Turkey 2000-2001	Argentina 2000-2005	Uruguay 2001-2003	Brazil 2002-2003
Pre-crisis debt indicators						
External debt						
as percentage of GDP	68	29	60	51	81	41
as percentage of exports	267	66	203	376	365	300
Short-term foreign debt (percent of reserves)	181	327	246	149	320	142
Public debt						
as percentage of GDP	67	41	53	45	38	73
as percentage of revenues	486	113	203	226	191	211
Foreign-currency debt (percent of total)	n/a	68	50	91	83	25
Pre-crisis memorandum items						
Current account deficit (percent of GDP)	-8.6	-3.1	-4.9	-3.1	-2.6	-1.7
Fiscal deficit	-4.1	-2.7	-10.4	-2.4	-4.4	-5.2
Primary balance	-1.3	-0.4	5.4	-1	-2.2	3.9
Interest payments (percent of revenues)	29	7	62	17	13	21
Exchange rate peg 1/	Yes 4/	No	Yes	Yes	Yes	No
Crisis characteristics						
Currency crisis	Yes	No	Yes	Yes	Yes	No
Banking crisis	Yes	No	Yes	Yes	Yes	No
Corporate financial crisis	Yes	No	Yes	Yes	No	No
Output fall	Very large	Modest	Large	Very large	Large	Small
Crisis resolution						
IMF and bilateral commitment (US$ bln)	0.0 5/	0.4 3/	33.8	23.1	2.7	35.1
IMF and bilateral commitment (percent of GDP)	0.0	0.8	17.0	8.1	14.5	6.9
Sovereign debt workout						
Bond restructuring	Yes	Yes	No	Yes	Yes	No
Bilateral loan restructuring	Yes	Yes	No	Not yet	No	No
Commercial loan restructuring	Yes	Yes 6/	Yes	n/a	n/a	Voluntary

Source: Roubini and Setser (2004a), IMF country reports, author's calculations.
n/a: not applicable. 1/ Any regime without free or managed float is subsumed as pegged, such as currency boards, fix or crawling pegs or bands. 2/ Additional to preexisting IMF support for the overall transition. 3/ Approximate extension of existing IMF program. 4/ Sucre floated in March 1999, i.e. shortly before default. 5/ No IMF program in place during crisis, IMF assistance commenced in 2000. 6/ Agreed upon in the terms of the Paris Club agreement.

2.2.1 Mexico 1994–1995

The Mexican crisis 1994–1995 became the first emerging market crisis descending from both current and capital account deficits. For this reason, IMF Managing Director Michel Camdessus earmarked the crisis as the "first of the twenty-first century". Besides "Peso Crisis" and "Tequila Crisis" (with respect to its consequences in the region), it is most accurately called the "Tesobono Crisis", referring to the crisis' roots in the unsustainable burden of short-term domestic notes linked to the US dollar. This burst of short-term debt emanated from the turbulent political circumstances in the election year. In response, the incumbent government expanded its fiscal expenditures and missed a realignment of the currency. Introduced under the outgoing administration of Carlos Salinas, the peso peg had contributed to low inflation and rapid economic growth. It was widely acknowledged, however, that the Mexican peso was overvalued and nurtured an economic bubble. The peg attracted large portfolio flows which usually prove very volatile and quickly evaporate upon first signs of a crisis. By the end of 1993, about 29% of all Mexican stocks and 79% of all bonds were held by foreign investors. In addition to this constellation, the country was facing a rollover of $30 billion of tesobonos in 1995.

With the new government not yet in office, a run on the peso started in November 1994. The hefty short-term consequences of the crisis were associated with the sudden and unexpected meltdown in December 1994 when the peso was floated; therefore called the "December Mistake". This resulted in an immediate devaluation of about 50%, widespread uncertainty and distrust within the financial and corporate sector. In 1995, GDP contracted by 6%.

The large concerted assistance (totalling about $50 billion and presenting a large overrun of traditional support granted by the IMF) was accompanied by an unexpected quick policy response by the new Mexican government under Ernesto Zedillo. The aid package was widely criticized as pure bail-out, and the U.S. was blamed for promoting too much multilateral help to fixing problems in their backyard. The rescue effort (tailored to overcome a liquidity, not a solvency crisis) was, nevertheless, a stunning success. An intervention among the recently privatized banks prevented a collapse of the banking system which had accumulated non-performing debt in response to expansionary credit concessions, low interest rates, and insufficient bank supervision during the boom years. Avoiding a default or restructuring of its sovereign bonds, Mexico regained access to the international capital markets in 1996. This was mainly due to the fact that solvency was not an issue. By 1997, the economy was booming. Today, the country is rated investment-grade.

2.2.2 The Asian Crisis 1996–1997

The quick and painless resolution of the Mexican crisis fueled investor optimism and increased worldwide capital flows into emerging markets. The sequence of crises in Thailand, Indonesia, South Korea, and (to a lesser extent) Malaysia came as a major blow from behind for investors. However, the crises did not have their seeds in excessive public borrowing, questioning the sovereigns' solvency. None of the countries affected encountered default or engaged in a sovereign bond restructuring.

Thailand was the first country to come under pressure in May 1997 when foreign exchange speculation put pressure on its baht-dollar peg. After floating the currency in July, the government negotiated a moderate aid package of $17 billion which was mainly used to restructure the banking sector. Both banks and private corporations had large foreign debts from excessive investment flows into the "Asian tigers". A wave of corporate bankruptcies caused widespread economic suffering among the Thai population. However, a well implemented stabilization program helped the economy to recover rapidly.

In October 1997, Indonesia also sought IMF assistance to support its sliding exchange rate. Despite a total commitment of $36 billion (of which about $14 billion was disbursed) the rupiah lost more than three-quarters of its value within one year. The government's endeavors to assist struggling banks and corporations resulted in a quick accumulation of Indonesia's public debt and forced it to seek restructuring agreements with the Paris and London Clubs. Investors and, to an inestimable degree, the population bore the crisis' burden since stabilization efforts were far less effective than in Thailand.

South Korea was also caught in a maelstrom: aware of the fragile banking system and a crusted corporate sector, a sellout of won assets started. An international aid package of $58 billion—larger than the Mexican bailout in both absolute and relative terms—quickly succeeded in stabilizing the most important economy in South-East Asia. Soon after the crisis the government was able to re-access the international capital markets with two bonds issued at moderate spreads of around 350 basis points.[26]

While interesting with regard to crisis mechanics, the Asian crisis proves only moderately relevant to this study as none of the affected

---

[26] See IMF Staff Country Report No. 00/11 (2000), p. 34.

countries were significant players in the sovereign bond markets. Still, it is useful to keep these events in mind as they unmasked a supposed economic miracle and reminded investors of the dangers of hidden imbalances and contagion.

2.2.3 Russia 1997–1998

In Russia, as well as many other transition countries, economic reform caused considerable declines in real GDP (such as 12% in 1991) and huge fiscal deficits (26% of GDP in 1991). After an eight-year recession, the Russian economy rebounded in 1997, supported by IMF programs installed in 1995 and 1996. While the programs introduced strict rules for the monetary policy, the fiscal deficit kept growing. The gap was refinanced in the newly established domestic market for short-term ruble bonds (the "GKO" and "OFZ"), mainly bought by domestic financial institutions and, in turn, refinanced abroad. This created an unsustainable mismatch over time. Notwithstanding this development, Russia succeeded in placing several eurobond and "MinFin" issues (domestic dollar denominated bonds), and reached agreements with the Paris and London Clubs. In the deal with the latter, the state-owned Vnesheconombank converted $31.8 billion of Soviet-era debt into principal notes ("Prins") and interest arrear notes ("IANs").

A slump in commodity prices along with investors' doubts about emerging market investments (as consequence of the Asian crisis 1997) caused a powerful external shock. Given a situation of internal political deadlock and a non-performing IMF program, the insolvent government was unable to stand the pressure in August 1998. The persistent fiscal deficit and the drain of reserves (from defending the ruble) forced the government to devalue the currency, as well as declare a moratorium on all public ruble debt and part of the foreign liabilities. Further political turmoil triggered a bank run, correctly forestalling a banking crisis as the banks carried the main burden of the GKO and OFZ default.

In December 1998, Russia defaulted on Prins and subsequently on IANs. Eurobonds and MinFins (except for the "MinFin III" issue) were not affected. The fact that the London Club did not call in its loans (which could cause Russia considerable legal trouble with creditors) helped to calm down the crisis. In 1999, another rescheduling with the Paris Club was settled, and the IMF resumed support. In February 2000, renegotiations with the London Club reached an agreement on the exchange of $22.2 billion of Prins, $6.8 billion of IANs and some $2.8 billion of past due interest. This restructuring offered substantial cash flow and debt relief, contributing to the country's solvency. Since then,

Russia has recovered astonishingly quickly. High commodity prices and sound fiscal and monetary policies helped to restore strong GDP growth since 2000. While the total external debt ratio declined steadily, the government became able to re-access foreign lending sources. Russia enjoys an investment-grade rating by Moody's since 2003 and by Standard & Poor's since 2005.

2.2.4 Brazil 1998–1999, 2002–2003

During most of the 1990s, Brazil followed a healthy economic path with sustainable fiscal policies but suffered from high inflation. The Real Plan of 1994, aiming at disinflation (among other objectives), is believed to have created the imbalances which lead to the crisis in 1998–1999. Deflationary fiscal and monetary policies turned out to increase the fiscal deficit, for instance through asymmetric indexation of revenues and expenditures. The crawling peg adjusted too slow to avoid an appreciation in real terms, causing the current account to turn into a deficit.

With adverse external developments like the Asian crisis, market pressure on the real increased. Invigorated by reelection, the Cardoso administration tried to maintain the peg as anchor for inflation expectations. Markets, however, believed in a significant overvaluation. The unsustainable situation was supported by an IMF program to restock reserves. However, market forces soon prevailed. Brazil floated the real in January 1999 after a reform of the peg resulted in record capital outflows of $14 billion within a few days. Although public debt, and especially external debt, quickly expanded in the run-up to the crisis, the burden was not believed to exceed the sustainable limit. However, with bond spreads reaching 1,500 basis points at the crisis' peak, it became impossible to access the international markets for liquidity. Instead, a rollover of short-term credits was negotiated with major international banks.

In the end, the output fall remained small and the banking sector got off lightly. Roubini and Setser (2004a) argue that the large amount of reserves spent on supporting an unsustainable exchange rate peg helped banks and firms to create hedges against a widely anticipated devaluation. Recovery was quick as growth resumed the following year, and IMF emergency loans were repaid ahead of schedule.

Still, the crisis left marks on the public budget. By offering dollar-linked debt during the currency crisis, the government created a currency mismatch on its own balance sheet. Its public debt ratio reached

70% in 2001. IMF assistance was readopted in the light of the Argentine crisis which, in the end, had less effect on Brazil than anticipated. Combined with concerns about the populist presidential candidate, Luiz Inácio Lula da Silva, and his platform plank not to repay foreign public debt, investor confidence slumped. Sovereign bond spreads widened to 2,000 basis points in autumn 2002. While solvency was clearly disputed, it is nevertheless hard to fit this bond market panic into the traditional scheme of solvency and liquidity crises. Lula's campaign against foreign creditors resembles more the antique fear about a sovereign's discretion to repudiate foreign creditors. Another major rescue package by the IMF, and an unexpectedly prudent economic policy by the new administration (lagging its previous rhetoric), succeeded in resolving most doubts. The international bond market was touched again for new issues in mid-2003. Even though high debt levels in the BB rated country persisted, external bond spreads declined to around 400 basis points in 2005.

2.2.5 Pakistan 1998–1999

Pakistan represents an example of a solvency crisis, but on a very small and less complex scale. The government's high indebtedness of up to 90% of GDP was mainly caused by a lax fiscal policy that wrote large deficits. Over time, the external position became less and less sustainable due to large current account deficits and declining worker remittances. This imbalance prompted fears of capital flight. The situation took a turn for the worse after the international community imposed economic sanctions in response to a series of nuclear tests in May 1998. To avoid further capital flight, the government imposed a deposit freeze. As a side effect, foreign exchange sources shut off and reserves started declining. This caused the government to slowly accumulate arrears, pushing the country closer to a complete default.

In January 1999, the situation stabilized when the IMF stepped in. In the same month, a debt relief agreement with the Paris Club stressed the requirement of a sovereign bond restructuring to ensure the comparability of treatment. After some initial hesitation (fueled by fears of reputational loss), a new (military) government put forward a voluntary exchange offer in November 1999. At that point, bond trading was already very illiquid, rendering a roll-over of the soon maturing bonds impossible. Under the terms of the offer, three eurobonds (maturing between December 1999 and February 2002) were eligible for the exchange. Even though no face value reduction arose from the exchange, Pakistan enjoyed a significant cash flow relief in the near-term since the duration was increased by more than three years. Due to the credibility of the default threat, the limited number of bondholders and the benign terms of the offer, the first comprehensive bond restructuring completed became a success. Resumed IMF and Paris Club support helped Pakistan to overcome the crisis by 2001. In 2004, Pakistan reaccessed the markets, issuing a 6 3/4% five-year bond, yielding about 6%.

2.2.6 Ecuador 1998–1999

The Ecuadoran case presents the first default on international sovereign bonds. Furthermore, it became the first default on already restructured Brady bonds, intended to be so-called "exit instruments" immune against future restructurings. This crisis presents a clear case of insolvency with the rigorous implementation of equal burden sharing.

Since Ecuador's late opening to international capital markets, it persistently ran into arrears with commercial and bilateral lenders. A large budget deficit (partly due to a military skirmish with Peru), corruption, and political instability did not help to establish a good reputation. Defying these odds, the government succeeded in placing two eurobond issues. By end of 1998, the situation worsened from a combination of factors. Damages from El Niño-induced storms, a commodity price drop, and capital market turmoil spilling over from Russia first caused severe liquidity problems among banks, and then exerted strong pressure on the pegged currency. After reserves could not withstand the capital flight, the sucre was floated. This in turn made many banks insolvent due to their large dollar liabilities. A deposit freeze was introduced, and a government agency—as part of a quickly introduced deposit guarantee—took over the most severely troubled banks.

In August 1999, the government held back payments on its Brady bonds, calling for the activation of the interest collateral. Creditors were appalled at the idea of a repeated default on a restructured instrument. In October 1999, the country also missed coupon payments on its eurobonds. In 2000, a new government dollarized the economy and agreed to IMF assistance. A restructuring offer was made in July 2000. It intended the conversion of all Brady and eurobonds into a new 30-year eurobond with step-up coupon which can be converted under a 35% face value reduction into a 12-year eurobond. Already restructured bonds were offered "Contingent Recovery Rights" as insurance against a repeated default before 2010. The increase of duration and the step-up feature contributed, together with a face value reduction, to the sustainability of the debt service burden. As a result, risk spreads decreased from well above 2,000 basis points to around 1,300. However, Ecuador did not access the international bond markets for several years and encountered serious refinancing problems in 2005 again.

2.2.7 Ukraine 1998–2000

After shedding its Soviet-era debt, Ukraine first touched the sovereign bond market in 1995 and piled up $1.6 billion in eurobond debt within three years. This changed the Ukrainian debt composition from loans (mainly owed to bilateral lenders) to domestic and external bonds bought by institutional investors in Europe and Asia. The debt burden turned fatal when external capital flows ran dry in the wake of the Russian crisis. Furthermore, the Russian devaluation caused a slump in Ukrainian exports, effectively increasing the pressure on the hryvnia. While multilateral assistance helped to avoid a further meltdown, the government decided for a considerable (but not drastic) devaluation. In the following, the government sought restructuring agreements of its bank debt (August 1998) and quasi-bond instruments (September 1998), and engaged in piecemeal reschedulings of single instruments judged as technical (but not de jure) default.[27] The augmentation of an existing bond became a market concern since this not only pushed down the price, but was also believed to expose existing investors to speculative interests because the additional bonds were mainly bought by vulture investors.

An unsatisfactory review of Ukraine's IMF program made future disbursements conditional on a comprehensive bond restructuring. Facing the maturity of all of its external bonds in 2000 and 2001, the government initiated an exchange of four eurobonds (and one bond owed to the Russian Gazprom corporation). The circumstances are therefore earmarked as a liquidity problem. To avoid holdouts, the government missed out due payments in January and February 2000 but launched its exchange offer during the grace period. The offer included the choice of two amortizing eurobonds denominated in euro and dollar. While maturities were extended to 2007, the offer did not ask for a reduction in face value and reimbursed accrued interest in cash. The exchange was facilitated by the limited number of institutional investors and the fact that three eurobonds were issued under Luxembourg law including collective action clauses. This contributed to a participation rate of 99%. Cash flow relief was mainly gained from increasing the average life

---

[27] For details see Sturzenegger (2002), pp. 35ff.

of the bonds. Despite considerably improving fundamentals in the following years, distress-like market spreads persisted the following year. In 2003, Ukraine resumed issuing sovereign bonds on the international debt markets.

2.2.8 Turkey 2000–2001

Ever since the beginning of the cold war, Turkey constituted an important strategic partner in international politics. This may have helped the country to obtain major rescue assistance during its long history of financial crises. After major crises in 1959, 1965, and 1979, Turkey started an economic stabilization program under a military rule in the 1980s. In 1989, Turkey liberalized its capital account to avoid crowding out domestic investors by the large demand for public credit. While direct investments proved healthy for the economy, the large and volatile portfolio flows made the country vulnerable to any slumps in global investors' confidence in emerging markets. Heavy inflows from abroad fueled economic growth but increasingly inflated the economy. Part of the boom was originated by large public investments which were refinanced abroad and came hand in hand with widespread corruption. As a result, Turkey became prone to characteristic boom-bust cycles.

After the emerging market jitters from 1997 and 1998, and an earthquake in 1999, overall economic conditions worsened, exposing the fragilities of the banking system as eight banks became insolvent. In December 1999, the government—backed by an IMF stand-by loan—installed a crawling peg for the Turkish lira to bring down inflation. Despite structural reforms and fiscal adjustments, the peg (preannounced to remain in place for only 18 months) failed to curtail inflation expectations. In response to the real appreciation, the trade deficit doubled. At the same time, the government refinanced the expanding fiscal deficit with short-term bills placed among domestic banks. Attracted by the high nominal rates, international reserves grew in absolute terms but covered only 50% of short-term external debt by the end of 2000.

The sudden collapse of investor confidence and starting capital flight in November 2001 occurred upon a number of events, notably the emergence of a banking scandal, political turmoil, and contagion from Argentina. Without the confidence of foreign investors, domestic banks were unable to roll over their debt, and bid on local markets for financing. A massive exit from the lira drained foreign reserves and contributed to a liquidity crunch in the domestic banks. While the situation calmed upon the agreement of a first $10.5 billion IMF aid package in December 2001, a political conflict over the peg fostered speculations.

In February 2001, finally, the peg was abandoned and the lira lost about one third of its value against the US dollar. Despite a revision of the IMF program and additional stand-by assistance in May 2001, large fiscal imbalances, political fragility, and high inflation made any policy efforts ineffective, causing an even weaker real sector performance than expected.

The events of 9/11 worsened the overall situation for emerging markets, but reinvigorated Turkey's strategic role in the Middle East. By the end of 2001, a new stand-by agreement provided additional funds. In the meanwhile, the country went through a typical deflationary period. Corporate insolvencies reduced the foreign exchange demand by defaulting on external debt. Despite a high public debt load (running close to 100% of GDP), continued IMF bailout packages, and sound fiscal management enabled the government to avoid a debt restructuring. The long-term sustainability of the debt is disputable. However, the government was able to overcome recent political irritations, like a break up of the coalition government in mid 2002 or the parliament's denial of the deployment of U.S. troops in Turkey for the Iraq war in 2003 (foregoing a large multi-year support package). With the EU accession talks underway, a convergence play helped pulling spreads down to little above 200 basis points in 2005.

2.2.9 Argentina 2000–2005

Argentina has produced a number of precedents, lately with its 2005 mega bond restructuring.[28] While counting as one of the wealthiest countries only one century ago, economic prosperity declined since the 1950s, mainly due to a heavily regulated economy left from the Perón era. Argentina became the first country to enjoy debt relief from the Paris Club in 1956. Under Carlos Saúl Menem and his economy minister Domingo Cavallo, a currency board with a one-to-one dollar peg ("convertibilidad") was introduced to fight hyperinflation in 1991. While inflation faltered and economic growth flourished, structural reforms (deregulation, trade liberalization, privatization) led to a larger public deficit after 1994. In 1993, the country received debt reduction under the Brady plan, granting the government access to the international capital markets. Future public deficits were financed by foreign debt issues.

---

[28] The Argentine crisis, and its relationship to the Fund, is subject to a large number of academic analyses, such as Mussa (2002), Daseking et al. (2005), Cline (2003), Perry and Serven (2003), and De la Torre et al. (2002).

Throughout the crises in Mexico and Asia, the Argentine economy appeared stable and served as role model for the IMF.[29] When the important Mercosur trade partner Brazil devalued 1999, Argentina was caught in an unsustainable situation. High domestic interest rates and an overvalued currency dragged the country into recession. Markets quickly deemed the external debt load unsustainable as its debt-to-export ratio increased to 400%. An economic program, announced under the name "blindaje" (armor plate) in December 2000, came along with the first rescue package from multilateral creditors valued at $20 billion. The impact of these measures quickly degraded due to the lack of both domestic support and external confidence.

First, a liquidity crisis unfolded. A lack of market access gave rise to the "megacanje", a mega bond swap in order to extend maturities in June 2001 when Argentina faced major rollover needs. A broad number of debt issues were eligible for exchange into one short local bond and three global bonds maturing in 2008, 2018, and 2031. While local pension funds were not left much of a choice, the tender was voluntary for other creditors. With a participation of $29 billion of debt (almost half of all eligible claims) the swap was a stunning success despite the conviction that the overall debt situation was unsustainable. However, the deal significantly increased the overall debt stock. Market confidence remained low and led to subsequent difficulties when provincial entities tried to roll over their debt.

Investors became increasingly convinced of Argentina's insolvency. Despite efforts to stabilize the situation, market spreads increased to 1,600 basis points. As capital flight intensified, the IMF reluctantly granted another credit line to prop up central bank reserves. After disappointing tax collection figures in September, the government sought support for voluntary debt relief in two stages. In the first stage, domestic bondholders were supposed to exchange their bonds into loan instruments under local legislation, guaranteed by tax revenues. These included reductions in coupons and a maturity extension. While the idea was to preserve the local financial institutions, creditors recognized the implied default threat and tendered almost all eligible claims in November. The segmentation of bondholders into domestic and foreign creditors was not perceived well by international investors. Rating agencies judged the transaction as technical default, making it impossible for the IMF to grant further support. Before announcing the second stage (which presented an exchange offer to foreign creditors), massive capital flight led to the collapse of two large banks. The policy re-

---

[29] The country received four IMF arrangements between 1992 and 2000.

sponse of a bank holiday and a deposit freeze ended in street riots, forcing the resignation of the Menem administration. Under the applause of the parliament, the new interim president Adolfo Rodríguez Saá announced the suspension of payments on all foreign bonded debt in December 2001.

Ousted after seven days into tenure, Eduardo Alberto Duhalde introduced a painful pesification of the banking sector, and first devalued and then untied the peso peg in the beginning of 2002. What followed was an economic crisis of historic proportions in Argentina. The peso overshot and temporarily lost three-quarters of its value relative to the dollar. Output fell by 20%, GDP faltered by 11%, inflation and unemployment shot up, and depositors lost much of their savings. The situation stabilized in the third quarter of 2002. In January 2003, the IMF resumed its support with a short-term stand-by arrangement. In 2003, GDP increased by almost 9% and unemployment was down by a quarter. Economic recovery and political stability under the new president Néstor Kirchner was remarkable.

Similar to his rough rhetoric with regard to multilateral agencies, the Kirchner administration took a hard stance towards the holders of the defaulted debt. After a failed sketch of a restructuring in 2002, a slightly improved exchange offer was launched to domestic and international bondholders in December 2004. Despite widespread resentment among the investors, the final participation rate reached 76%, achieving substantial debt relief for Argentina. Total sovereign debt fell by one-third to around 81% of GDP, converting part of the external debt into local currency. Regardless of the uncertainty about hold-outs and a wave of lawsuits, the government is expected to re-access international bond markets soon. After the swap, the country was assigned a B- rating by S&P, and bond spreads tightened to less than 400 basis points.

2.2.10 Uruguay 2001-2003

Although severe contagion became a concern after the Argentine collapse, Uruguay was virtually the only country that was seriously hit. Stemming from a combination of factors, large capital outflows occurred in 2002 and 2003. Even though an IMF program was in place and helped to calm down investor sentiment, the run made parts of the banking sector insolvent and triggered a devaluation of the peso.

Since the country's once notable credit standing on the international markets created an overhang of external debt, an immediate liquidity problem arose when the government faced a restructuring of the banking sector. Although enjoying an investment-grade rating only one year before, Uruguay launched an extensive exchange offer in the spring of 2003. The restructuring, supported by the IMF, covered 46 domestic and 18 international bonds with a total principal of $5.4 billion which is almost all of Uruguay's bond debt and about half of its total public sector debt. The exchange was designed to significantly reduce financing needs through the international capital markets during the following years. In addition, the exchange sought to improve debt sustainability in the medium term while condensing the variety of bonds outstanding in some larger issues. These objective were achieved by offering an exchange, mostly at par, with two options. The maturity extension option included a new bond with an extended maturity of generally five years. The benchmark bond option offered an exchange into larger and more liquid bonds with a maturity of 30 years. This helped to gradually reduce par spreads, declining from above 1,500 basis points in 2002 to below 400 basis points in 2005. However, it is still disputed whether the exchange was appropriate given the considerable external debt burden at that time. Uruguay regained a B rating and issued a $150 million note with a 17 3/4% coupon in August 2004.

2.2.11 Moldova 2002

The Moldovan case represents a clear-cut liquidity crisis of very small dimensions. The republic gained independence from the former Soviet Union in 1991, but has remained under strong Russian influence since. Despite a favorable macroeconomic outlook, liquidity constraints arose shortly before the maturity of its only eurobond in June 2002. Supported by the fact that 78% of the outstanding bonds were held by one investor, the government engaged in restructuring negotiations. These were facilitated by the activation of a collective action clause which required a 75% majority vote. The exchange offer in October 2002 included a cash payment of 10% of the outstanding principal and a new dollar denominated bond maturing in 2009. The new bond started amortizing in 2003 in order to avoid another liquidity squeeze in the long term. During this transaction, an IMF program was in place and the transaction was welcomed by the Fund and other multilateral agencies.

2.2.12 The Caribbean Restructurings 2005–2006

A number of Caribbean countries suffered considerably from natural disasters and a decline in tourism after 2001. IMF programs were designed to resolve these problems. The programs envisioned bond restructurings that were comparatively small. The following briefly reviews the transactions in Dominica and the Dominican Republic.[30]

Dominica suffered from a row of serious, permanent external shocks, causing public debt to expand quickly and shutting off access to foreign capital. The subsequent bond restructuring presented a characteristic case of a preemptive exchange apart from the fact that its two bonds were subject to a legal dispute. Since the original bonds were horizontally stripped into zero coupons and sold to a wide range of regional investors, the main challenge of the transaction was gaining a critical acceptance rate. For this reason, the exchange remained open for the a prolonged time in 2004. The long duration of the new bonds and a small face value reduction not only resolved liquidity problems, but also improved the sovereign's solvency.

In the Dominican Republic, the economy enjoyed rapid economic growth during the second half of the 1990s, after the Balaguer regime fell. In 2001, the global downturn and a decline in tourism curtailed growth and unemployment shot up. A strong policy response and a reliable relationship with the IMF supported market confidence, enabling the country to issue an international sovereign bond in the same year. In 2003, a major banking scandal triggered market pressure on the peso and a slump in investor confidence. The subsequent economic program came at large costs to public finances (increasing the public sector debt from 26% in 2002 to 45% of GDP in 2003) and was financed through another international sovereign bond. Besides decisive structural reforms with strong IMF support, a bond restructuring was initiated to facilitate cash flow relief for 2005 and 2006. The offer was finally launched in April 2005 when the government was already in arrears with a coupon payment and part of its commercial debt. The offer aimed at extending solely the maturities in order to comply with the comparability provision of the Paris Club. In the aftermath, spreads fell

---

[30] At the time this chapter was written, Grenada also engaged in a restructuring of two international sovereign bonds. Facing already severe economic problems, Hurricane Ivan caused damages measuring 200% of GDP in 2004. With public debt increasing to 130% of GDP and public revenues sharply contracting, the government missed payments on the bonds but stayed current on commercial and bilateral debt obligations. The conditions of the exchange offer were outlined in September 2005.

from above 1,500 basis points in summer 2004 to less than 450 basis points one year later when the country received a B rating by S&P.

2.2.13 Outlook

What can we learn from the past? Historical experience has shown several ways of sovereign lending and crisis resolution. Whereas sovereign defaults have been decreasing for four decades since World War II, the wave of defaults in the 1980s solely affected bank loans. Even in the 1990s, with global bonds not yet playing a pivotal role in public finance, sovereign bonds were able to escape default—the Mexican liquidity crisis in 1994–1995 was averted by generous international support, preventing a precedence for a "modern" sovereign default from happening. Apart from smaller incidences, the restructurings of Ecuador and Pakistan in 1999 became the archetype of modern restructurings, infamously followed by the exchange of defaulted bonds from Argentina.

The type of crisis predominant since the 1990s suggests (regardless of all its idiosyncratic features) common causes, such as maturity and currency mismatches, fragile banking systems, and exchange rate overvaluation. Unfortunately, this has not made it easier to foresee the timing and extent of a crisis which may affect the banking and corporate sector, the exchange rate, and sovereign debt. In any case, crisis resolution always involves significant expenses for the government. Given a sound emergency plan to contain the crisis, the provision of liquidity plays an invaluable role in crisis resolution. As markets are usually unable to perform this task, the importance of emergency assistance provided by the IMF should be undisputed. However, if a government's expenses for crisis containment create an unsustainable debt burden, the liquidity problem may tip over into a solvency problem. In the latter case, exiguous multilateral support may look like "bailing out investors", which pull out their money in anticipation of large write-offs.

However, making a distinction between illiquidity and insolvency at the onset of a crisis is difficult, and political considerations have undoubtedly strengthened the case for prolonged bi- and multilateral support even when it seemed ill-advised. With emergency loans adding a senior tranche to the external burden of public debt, this might accelerate the transition from a liquidity to a solvency crisis. In the latter case, private creditors will inevitably be forced to provide long-term debt relief. Whether sovereign bond investors are asked to share in mainly depends on the proportion of bonds relative to other debt instruments, traditionally referred to as "de minimis rule". The new paradigm of private sector involvement largely avoids making an ex ante distinction between illiquidity and insolvency. In any case of significant commercial loan or bond debt, bi- and multilateral agencies will call for an investor bail-in. While the restructurings of Pakistan and Ecuador paved the way for such transactions, the workouts in Ukraine, Uruguay, and the Caribbean countries make a case for expecting more sovereign bond restructurings to become part of future crisis workouts.

This being said, crises might still unfold despite deepened global integration and the promising development of the developing country class in recent years. While resorting to speculations is misplaced here, some countries show certain vulnerabilities with respect to sovereign indebtedness. The continued struggle of Ecuador to refinance its debt is an obvious example. Unusually abundant liquidity on international capital markets has recently reduced spreads to record lows. As global monetary conditions tighten and investors apply closer scrutiny, a reversal of this development might reveal unforeseen liquidity and solvency problems. Similar cycles have been present in the past, and their reoccurrence should be no surprise.

2.3 The Theory of Sovereign Lending and Default

With past experiences in mind, the following sections explain how the scientific literature has formed theoretical models and empirically analyzed the mechanics of sovereign lending and crisis resolution.

Borrowing and lending enables the intertemporal shift of consumption and investment. Thus, public lending can help for consumption smoothing (Barro (1979, 1995)), or is used to increase public investment (such as infrastructure or education), exploiting higher rates of return and promoting growth and welfare (Eaton (1993)). From this view of the neoclassical growth theory, sovereign bonds are simply one instrument among many to facilitate capital flows, circumventing the "original sin" problem by borrowing in foreign currency (Eichengreen, Hausmann and Panizza (2003), Özmen and Arinsoy (2005)) and exploiting competitive lending on the international capital markets. The welfare effects are constrained by the necessity of repayment from tax revenues and the "crowding out" problem with regard to private borrowers. While the latter plays an increasingly important role as debt flows to private entities in emerging markets expand, the former has already lead to instabilities on many occasions. Where the public budget becomes difficult to balance, fiscal policy to maintain debt service tends to become pro-cyclical (rather than counter-cyclical), further reducing the welfare effect of public borrowing and creating incentives for debt repudiation.

The following reviews the current theoretical literature on sovereign lending which can be divided into three areas: Firstly, the theory of lending explores why sovereign debt exists despite the lack of an orderly bankruptcy procedure for sovereigns. Secondly, the crisis literature researches causes and consequences of financial crises in emerging markets and contagion effects. A final area of study develops approaches for the containment and resolution of crises and gives recommendations on the role of the international financial institutions.

2.3.1 The Theory of Lending

Theoretical models of sovereign debt strive to determine the optimal debt level for a sovereign and attempt to explain under which conditions a sovereign would ever repay. As there is not any enforcement mechanism to ensure repayment, these models utilize a set of benefits or penalties to foster repayment, such as exclusion from capital markets, loss in output, or international trade retaliation.[31]

Consumption smoothing is the central motive in equilibrium models while maintaining a "good" reputation on international capital markets serves as the enforcement mechanism (posing the threat of financial autarky to a sovereign debtor).[32] A reputational equilibrium is not achieved in case where capital investment is the motive for sovereign lending. This sort of lending is driven by a higher-than-average return on investment. As long as the higher returns persist, there is an incentive to take on more loans. By backward induction, no lending will be made in the first place (Eaton et al. (1986), Rosenthal (1991)).

Eaton and Gersovitz (1981) present an early example of a reputation model with consumption smoothing and complete information. A stationary equilibrium level of borrowing can be established by a concave utility function and under the assumption that debt repudiation bars the country from obtaining any loans in the future.[33] Loans are taken

---

[31] On output losses, see Barro (2001), Calvo (2000), Cerra and Saxena (2005), Cohen (1992), and Dooley (2000). Rose (2005) finds an 8% downturn in bilateral trade, persistent for 15 years, after Paris Club renegotiations. See Conklin (1998) for anecdotal evidence of currency transfer embargoes in the medieval period of King Philip II of Spain.

[32] See Eaton and Fernandez (1995), pp. 14ff, for a review of different standpoints on reputation.

[33] Related models under the same assumption are Grossmann and van Huyck (1988), Atkeson (1991), and Cole et al. (1995).

out solely for consumption smoothing and are to be repaid quickly; circumstances which are rarely observed in practice. Furthermore, instead of infinite condemnation to financial autarky, some agreement between lender and borrower could lead to a mutually beneficial new equilibrium.

Addressing these shortcomings, Bulow and Rogoff (1989a) allow for partial default and ongoing renegotiation of debt contracts, and incorporate stochastic interest rates and output. In their model, gains from trade serve as collateral, and a loan could be characterized as Nash-bargained insurance premium paid by the exporting country (the debtor) to a monopolistic party (the creditor) for not interfering with the debtor's trade ("trade of goods for sanctions"). The conclusion of Bulow and Rogoff (1989b), that reputation alone is deemed an insufficient incentive for repayment, has been challenged by a number of authors.[34] Following the concept of reputation, Kletzer and Wright (2000) show that a debt moratorium serves as sufficient punishment, which is unwound upon successful renegotiations, and that compliance with such a punishment mechanism is self-enforcing (by "cheat the cheater"-patterns). This adds to the existing literature by foregoing the implicit assumption of some third party enforcing commitments, such as lender seniority or monopoly rights in trade.

Up to now, this academic work yields insights into the equilibrium level of foreign sovereign debt, the so called credit ceiling of a debtor, while the cost of borrowing is exogenously specified. Pricing claims subject to sovereign default, however, calls for models which help determine the marginal cost of additional borrowing. By modeling debt to induce a drag on GDP, Kulatilaka and Marcus (1987) recognize the existence of a default premium over the risk-free rate and determine the optimal first passage time for default when reputation is at risk. Gibson and Sundaresan (1999) characterize sovereign spreads when exports serve as collateral, determining the recovery value. Default followed by a reduction in economic growth and a restructuring of foreign obligations is modeled by Westphalen (2001). His model facilitates the determination of the endogenous default boundary and the recovery rate, from which sovereign credit spreads can be calculated.

This literature differs from earlier studies of sovereign lending by incorporating stylized facts. Some authors have gone even further. The existence of liquidity crises, as trigger of restructurings, is examined in the literature on self-fulfilling credit runs (Chang and Velasco (2000),

---

[34] See Kletzer and Wright (2000), Cole and Kehoe (1996a, 2000), Chari and Kehoe (1993), and Kehoe and Levine (1993).

Cole and Kehoe (1996a, 2000), Detragiache and Spilimbergo (2001), and others). Alfaro and Kanczuk (2005) make an argument towards delaying default in order to maintain reputation and sending strong signals despite default-prone fundamentals, a scenario called "muddling through" which is observed frequently in practice.

By explaining the existence of sovereign lending and risk spreads, these studies have helped shape the financial architecture of the sovereign bond market. The design of sovereign lending instruments and the discussion about sovereign restructuring draws largely on the above fundamental concepts. However, empirical calibrations have proven difficult given the lack of well defined and regularly reported proxies for variables such as country wealth and reputational costs. The main contribution of this literature lies therefore in the application of game theory and asymmetric information to sovereign indebtedness, helping to develop an efficient sovereign debt market and suggesting mutually beneficial approaches to sovereign debt workouts.

2.3.2 Crisis Literature

The recurring appearance of emerging country crises has given rise to its own strand of literature, which begins with models on currency crises and subsequently shifts to models incorporating banking and sovereign debt crises. The understanding of the fundamental causes and transmission channels of these crises has helped to develop many of the empirical models of sovereign risk. This literature advanced in lock-step with the changing circumstances of crises in the last few decades, and has been characterized by first, second, and third generation models.

The first generation of model focuses on currency crises which emanate from fundamental macroeconomic weaknesses. The sovereign assumes domestic debt under a fixed exchange rate regime with international capital mobility. These circumstances were typical of the 1980s debt crises and became the basis for IMF programs until 1993. In academic models, foreign reserves are used to cover fiscal deficits, eventually making it impossible to credibly defend an exchange rate peg against speculative attacks (Krugman (1979), Flood and Garber (1984)). Later, first generation models were extended to include current account imbalances and real exchange rate misalignments.

The unpredictability of crises suggests the existence of multiple equilibria when fundamentals remain unchanged (like investor runs in a prisoners' dilemma setting). Second generation models take into account the political economy of maintaining currency pegs. The relevant background is provided by the European Monetary Union crisis in 1992 and the Mexican peso crisis in 1994–1995. Besides fundamentals, these models allow self-fulfilling investor expectations to enter the picture, stressing the role of speculative capital flows and short-term access to foreign reserves (Cole and Kehoe (1996b), Detragiache (1996), Drazen and Masson (1994), Obstfeld (1994), Sachs et al. (1996)).

After the Asian crisis, third generation models presented a more comprehensive picture, linking sovereign debt crises, currency crises, and corporate and banking crises. The transmission channels of such crises are mostly found in the financial sector while investor behavior is influenced by the existence of implicit guarantees, such as the pledge to maintain the exchange rate, or the bail-out of the banking system. The term "twin crises" (the joint occurrence of banking and currency crises) was coined by Kaminsky and Reinhart (1999), and other authors associated it with bank runs (Chang and Velasco (1999, 2001)) and credit boom-bust-cycles (Corsetti et al. (1999), Schneider and Tornell (2004)).

This advancement blurred the lines between corporate, banking and sovereign payment crises, and stressed the importance of public debt dynamics, as Roubini and Setser (2004a), pp. 26f, point out:

> "The systemic collapse of the corporate sector typically bankrupted the banking sector, and the cost of saving the banking system increased the government's own debt. In some crises, domestic banks borrowed from abroad to purchase the government's domestic debt, blurring the lines both between a domestic and an external crisis, and between a sovereign and a banking crisis. Most sovereign debt crises contaminate the banking system in some way (often because banks hold large amounts of government debt) and trigger large falls in the currency value that create payment problems for many firms."

A further layer of complications is added through the presence of cross-country spill-over effects, giving rise to studies of contagion.[35] International linkages and globalization is thought to have increased the presence of systemic risk, especially after the Russian crisis, but the lack of large-scale contagion after the Argentine crisis has calmed down the discussion.[36]

Fundamental contagion describes how shocks from one country are transmitted to others by means of macro linkages such as international

---

[35] Claessens and Forbes (2001) present an overview of this literature.

[36] An assessment of the Asian crisis with regard to contagion is presented by Baig and Goldfajn (1999).

trade and competitive devaluations.[37] Financial contagion models look at the reaction of the international financial markets. A number of crises were followed by a global decline in investors' appetite for emerging market investments, such as the slump in capital flows after the Asian and Russian crises, or the widening of risk spreads in Latin America during Argentina's struggle. While some might explain such reaction with behavioral arguments (e.g., investor herding in response to asymmetric information, see Calvo and Mendoza (1999)), another interpretation is that each crisis offers new insights which can lead to a reassessment of default risk ("wake-up call", see Pericoli and Sbracia (2003)). The relationship between empirical risk spreads and fundamental indicators of default risk is therefore subject to permanent change, posing a challenge to fundamental models of credit spreads.

2.3.3 The Literature on the IMF's Role

Connected with the literature on crises is the question of how to avoid, address, and resolve crises. The role of international financial institutions, particularly the IMF, has often been placed at the center, with important implications for sovereign debt. While recent proposals for sovereign debt workouts are discussed in Chap. 3, the following reviews the theory of catalytic finance.

In 1944, the Bretton Woods agreement included the founding of the International Monetary Fund and the International Bank for Reconstruction and Development (IBRD, the core institution within the World Bank Group). Later, further multilateral development banks formed around the initial institutions.[38] These new players became the catalysts for developing country lending in the period after World War II. While the World Bank and the regional development banks focus on private initiatives and project financing, the general financial architecture is influenced primarily by the IMF. Additionally, the IMF tries to act as the central institution to enforce commitments (which is implicitly assumed by some reputation models of sovereign lending) through the provision of financial assistance in crises. The assistance

---

[37] Corsetti et al. (2000) present a theoretical model. These channels were at work, for instance, in Argentina (in response to Brazil's devaluation) and Uruguay (in turn after Argentina's devaluation).

[38] Such as the Inter-American Development Bank (founded in 1959), the African Development Bank (founded in 1964), the Asian Development Bank (founded in 1966), and the European Bank for Reconstruction and Development (founded in 1991), among a number of smaller ones.

typically consists of an emergency loan package from several multilateral institutions which is monitored by the IMF and enjoys de facto highest seniority.

The theoretical literature focuses on two effects of this IMF "catalytic financing". First, the participation in an IMF program is seen as a signal in a principal-agent relationship of borrower and lender when the quality of macroeconomic policy is unknown. As IMF programs impose a cost on governments to increase their efforts and implement "good policies", entering a program serves as a screening device for private investors (Thomas and Marchesi (1999), Killick (1997)). Additionally, IMF surveillance, an important part of every IMF program, creates public information, supporting markets' informational efficiency (Fama (1970)). However, the efficiency of the IMF-induced macroeconomic adjustment itself is controversial, subject to dispute in the empirical literature.

Second, as a response to self-fulfilling creditor runs, IMF support is widely seen as a "liquidity insurance" by a lender of last resort.[39] While the previous aspect of IMF lending stresses measures which improve the debtor country's solvency and transparency, IMF support to prevent creditor runs consists solely of the unconditional provision of additional short-term credit lines. Based on models of domestic deposit insurance (Bryant (1980), Diamond and Dybvig (1983)), an international analogy can be established under some caveats (Goodhart and Huang (2000), Giannini (1999)). Similar to the theory of corporate finance, short-term debt (that causes possible roll-over problems, culminating in a liquidity crisis) is interpreted as a disciplining mechanism (Jeanne and Wyplosz (2001), Kumar et al. (2000)). While much of the past controversy focused on the optimal extent of liquidity provision under resource constraints, the IMF's present policy aims at preserving the vital financial architecture of a country and avoids financing speculative capital outflows, which is in line with theory (Jeanne and Wyplosz (2001)). In contrast to the political discussion, the theoretical literature has recognized that the key question is not whether to supply full (and unlimited) liquidity support or not, but where to draw the line in a trade-off between containing self-fulfilling runs and creating moral hazard, a classical problem of insurance providers.[40] The moral hazard problem may thereby be present on both sides. Foreign creditors, on

---

[39] The insurance consists of the provision of liquidity, not the subsidy element of below-market rate IMF loans since countries can get completely barred from financial markets, unable to borrow at any rate.

[40] See Roubini and Setser (2004a), pp. 74ff.

the one hand, may see an incentive to over-lend in expectation of IMF bail-outs that enable them to withdraw their funds. On the other hand, debtor governments may also lean towards excessive borrowing instead of pursuing economically sound, but potentially painful politics.

The theoretical literature has yet to produce consensus on this problem. Depending on the model setup, however, there is a case for efficient interventions given that these are ex ante not certain (Jeanne (2000), Ghosal and Miller (2003)) and remain partial (i.e., they leave some costs to creditors and debtors, see Corsetti et al. (2005)). This will, however, never completely rule out the possibility of liquidity crises occurring (despite sound fundamentals) as investors are risk averse (or even maybe irrational), informational asymmetries persist (Ghosal and Miller (2003)), and random shocks occur (Spiegel (2005)). Liquidity insurance offers help only in the presence of a liquidity crisis, but not in a solvency crisis (which usually requires a restructuring with substantial write-offs). Since these two situations are often difficult to distinguish, catalytic financing is identified to work best when fundamentals are sound (Corsetti et al. (2005)). Given these distortions and shortcomings, Kumar et al. (2000) and Miller and Zhang (2000) suggest that sovereign payment suspensions alone ("standstills", the equivalent of a "bank holiday") can work equally well in preventing creditor runs. This gives rise for a standardized procedure for workouts, a discussion pursued in Chap. 3. The literature does not yet offer a systematic comparison of both liquidity insurance and standstills.

Both the facts that creditor runs are hard to predict and IMF support cannot be taken for granted add to the uncertainty surrounding the valuation of sovereign bonds. It is difficult to evaluate how these measures influence the likelihood of a crisis and its cost to bondholders, and how all this combines into the ex ante perception of sovereign riskiness. The empirical literature, reviewed in the following, gives a first glimpse of how IMF action influences capital flows and risk spreads.

2.4 Empirical Evidence

Empirical applications of the above theoretical models are often troublesome and lead to unsatisfactory results. Foregoing much of the uncertainty surrounding the cost of default, empirical models strive to link macroeconomic fundamentals to sovereign lending. The topic most relevant to this study lies in the determinants of sovereign bond spreads. Before turning to this in Section 2.4.4, the earlier literature with related linkages will be discussed, such as the determinants of crises, the effect of IMF intervention, and sovereign ratings.

2.4.1 Determinants of Crises

In the empirical literature, many attempts have been made to explain past sovereign debt crises and use the result for prediction models, such as "Early Warning Systems". Most of the preceding literature focused on general balance-of-payment crises (Berg and Pattillo (1999), Berg et al. (1999)). Among the extensive empirical literature on currency crises, Reinhart (2002) shows that in four out of five debt crises a currency crisis preceded, exemplifying the interlinked nature of all types of crises. This is also highlighted by Allen et al. (2002), illustrating balance-sheet mismatches in different parts of an economy that can trigger a financial crisis.

Table 2.3 offers a synopsis of the literature which mainly strives to explain the occurrence of sovereign debt crises.[41] Thereby, a crisis is mostly defined by a rescheduling, arrears on principal or interest, or a default rating. By including IMF upper-tranche arrangements (or "aid packages") as a crisis event, some studies avoid dropping incipient debt crisis from the sample which were only averted by exceptional multilateral financial support. The dichotomous crisis variable is then related to fundamental factors by means of probit or logit regressions. As most studies focus on the 1970s and 80s, the findings reflect debt servicing difficulties mainly in connection with loans, but not bonds. Among the studies presented, Manasse et al. (2003) and Manasse and Roubini (2005) can be ranked currently as the most advanced approach. Their work applies an innovative statistical technique to identify non-linear interactions of crisis indicators which improves the predictive power and better distinguishes between different types of crises, requiring different policy responses. Their results highlight certain crisis-prone combinations of indicators, visualized in an empirically calibrated tree, and suggesting four types of countries: those prone to liquidity risk, solvency risk, macro-fundamental risks, and relatively safe countries.

---

[41] Related studies with divergent focus, not included in the table, are Berg and Sachs (1988), Lanoie and Lemarbre (1996), and Lloyd-Ellis et al. (1989, 1990).

Table 2.3. Binary studies of sovereign debt crises

	(1)	(2)	(3)	(4)	(5)	(6)	(7)	(8)	(9)	(10)	(11)	(12)	(13)	(14)	(15)	(16)	(17)	(18)
No. of countries	60	26	62	93	5	79	109	75	33	19	93	35	52	80	138	78	25	78
Start	1969	1960	1971	1971	1961	1971	1970	1970	1971	1971	1970	1980	1980	1980	1976	1971	1970	1984
End	1982	1968	1982	1982	1982	1982	1986	1985	1984	1986	1987	1990	1988	1990	1993	1998	2001	1997
Solvency indicators																		
External debt load 1/			+					+										+
Liquidity indicators																		
Debt service ratio 2/	+	+		+		+	+		+	+	+	+	+	+	+	+	+	
Reserves 3/	–	–		–	–		–		–	–	–	–	–	–		–		
Short-term debt ratio		+6/											+					
Other indicators																		
GDP growth 4/										–							–	
Per capita income				–				–	–									
Current account balance 1/	–		–			–										–		–
Inflation												+						+
Volatility 5/												7/					+	+8/
Exchange rate overvaluation																	+8/	+
Global interest rate								+									+	+

Sources: Peter (2002) and author's compilation. Sorted by year of publication. Positive and negative signs indicate significant regression coefficients. Numbers correspond to following studies: (1) Cline (1984), (2) Frank Jr. and Cline (1971), (3) Callier (1985), (4) McFadden et al. (1985), (5) Citron and Nickelsburg (1987), (6) Hajivassiliou (1987), (7) Hajivassiliou (1989), (8) Lee (1991), (9) Balkan (1992), (10) Li (1992), (11) Hajivassiliou (1994), (12) Odedokun (1995), (13) Marashaden (1997), (14) Rivoli and Brewer (1997), (15) Aylward and Thorne (1998), (16) Detragiache and Spilimbergo (2001), (17) Catao and Sutton (2002), (18) Peter (2002). 1/ As ratio of GDP or exports. 2/ Interest and principal payments as ratio of exports or GDP. 3/ As ratio of imports or total external debt. 4/ Real growth or real per capita growth. 5/ Volatility of measures such as exports (Frank Jr. and Cline (1971)), GNP-per-capita growth (Peter (2002)) or terms of trade (Catao and Sutton (2002)). 6/ Average remaining maturity. 7/ Positive and significant coefficient for the rate of real depreciation. 8/ Squared percentage deviation of real exchange from long-run trend.

Although this part of the literature does not consider the marginal cost of borrowing as the dependent variable, the results offer some guidelines for identifying common indicators of crises. It has to be borne in mind that the choice of the explanatory variables underly severe constraints in terms of data availability and frequency, which partly explains the omission of obvious indicators such as the public deficit. It is, however, difficult to infer from these findings the kind of default measures used in bond pricing, such as the default probability and recovery rate. Besides the common caveats of empirical determinants of crises, two more shortcomings are inherent in the above studies. First, the models cannot determine the political will (by the debtor itself or the IMF) to avert a crisis or soften the terms of a restructuring. Second, a study of a dichotomous event variable is hardly transferable into a continuous and forward-looking measure like bond spreads.

2.4.2 The Effect of IMF Involvement

A large number of studies is dedicated to determining the effect of IMF crisis lending on capital inflows and risk spreads. The success of IMF programs is highly disputed in the literature, but remains a controversy beyond the scope of this study.[42] The main question is, however, whether IMF intervention creates systemic effects on sovereign bond prices and how these effects evolved over time. For this reason, the following looks at the empirical findings on the effectiveness of catalytic finance versus the existence of adjacent moral hazard problems.

The evidence is, unfortunately, mixed due to the small number of clear bail-out cases, continuously changing policies, and a reverse causality problem as IMF loans are mostly directed to already struggling countries. Cottarelli and Giannini (2002) offer an overview of the empirical studies and point out general empirical hurdles. The effects of catalytic finance, i.e. an increase in capital inflows in response to a signal like an IMF program, are generally found to be weak (e.g., Bird and Rowlands (1997, 2000)). In case studies on general determinants of all kinds of private capital flows, catalytic effects appear to play only a subordinate role (e.g., Hajivassiliou (1987)).

Disentangling debtor and creditor moral hazard from risk spreads is difficult. Indeed, the link between macroeconomic and lending data suggests some weak evidence for debtor moral hazard (Dreher and Vaubel

---

[42] On this topic see, for instance, Barro and Lee (2001), Bird et al. (2004), Dicks-Mireaux et al. (2000), Edwards (1989), Haque and Khan (1998), Hutchison (2001), and Prezeworski and Vreeland (2000).

(2004), Gai and Taylor (2004)). As crises, even given IMF support, bear high costs to the government, it is more convincing to argue that the incentive to incur risky policies remains low (Roubini and Setser (2004a), pp. 107f). The empirical evidence on creditor moral hazard, at least when measured by linking IMF support to capital flows and spreads, underly the general criticism formulated by Jeanne and Zettelmeyer (2004):[43] As IMF presence is supposed to be efficient, higher capital flows or lower spreads just prove a necessary (but not sufficient) condition for moral hazard as these effects are also the clear objectives of the IMF. As long as IMF lending does not contain a large subsidy element, moral hazard cannot be present per definition (as the "Mussa argument" claims).[44] Since the subsidy contained in IMF lending is believed to be very small in emerging markets (i.e. apart from HIPC countries)—considering the fact that IMF debt is de facto senior and influences domestic policy via the conditionality channel (Jeanne and Zettelmeyer (2001), Zettelmeyer and Joshi (2005))—, this leaves little indication for creditor moral hazard.

Accordingly, empirical evidence is mixed. Risk spreads of commercial loans usually widen when an IMF program is adopted (Eichengreen and Mody (2000a), Haldane (1999), Ozler (1993)). A signal of good policies (leading to lower spreads) is found to work best with precautionary IMF programs, i.e. before a crisis-like deterioration of fundamentals (Mody and Saravia (2003)). The opposite impact is found for repeated use of IMF programs, constituting a similar effect as a reputation loss due to earlier debt servicing problems. The coincidence of IMF programs and private sector involvement during the Brady deal era (e.g., Marchesi (2003)), however, does also not necessarily prove a causality.

The changing nature of sovereign debt markets and associated reforms of the IMF's policy make it difficult to reach a final conclusion. Event studies and simple stylized facts may be more conclusive but also underly some econometric caveats. Lane and Phillips (2000) and Zhang (1999) do not identify significantly reduced sovereign spreads after the Mexican bailout which could have induced creditor moral hazard. The surprising default of Russia (a country considered as too strategically important to fail) has undoubtedly lead to a large increase in global sovereign spreads, which is regarded as evidence of moral hazard (Dell'Ariccia et al. (2002)). Anecdotal evidence suggests

---

[43] This caveat is already mentioned (but rarely addressed) in some studies, like Dell'Ariccia et al. (2002), Mody and Saravia (2003), or Lane and Phillips (2000).
[44] See Jeanne and Zettelmeyer (2004)., pp. 3f.

the existence of a "moral hazard play" in Russia, and similarities might be applicable to Brazil in 1998 and to Turkey (see Lane and Phillips (2000) and Roubini and Setser (2004a), pp. 106f). Analogous to what is believed to be the impact of measures that make debt restructurings easier (such as collective action clauses, see Chap. 3), Dell'Ariccia et al. (2002) find that countries with weak fundamentals suffered most from the Russian "nonbailout". The resumption of large-scale financial support in recent years (Argentina, Uruguay, Brazil), coupled with the withdrawal of support and subsequent default in Russia and Argentina, might have lent little reason to believe that creditor moral hazard plays a significant role today. In fact, as Chap. 3 argues, the inclusion of sovereign bonds in recent debt workouts (as part of the private sector involvement philosophy) implies that bond investors are unlikely to be completely bailed out by the IMF in any future crisis. Rather, IMF involvement in debt crises might catalyze a sovereign bond restructuring where the loss rate, given action is initiated from the outset, remains marginal. When splitting bond spreads into default intensities and recovery rates in Chaps. 5 and 6, the results might help to illustrate this evolution.

2.4.3 Determinants of Ratings

Ratings serve as a benchmark when comparing country risk cross sectionally. The transition of ratings, especially between investment and speculative grade, often motivate large portfolio rebalancings, causing immediate reactions on bond spreads. The rating methodology also exerts some pressure on the style and conditions of restructuring deals, as will be pointed out in Chap. 3.

While not the main focus of this study, it is therefore worth to briefly review the empirical literature on sovereign ratings. The dominant players among country risk ratings, relevant to foreign currency sovereign bonds, are the letter grade ratings provided by Standard & Poor's and Moody's. Other agencies or providers of related ratings (such as the Institutional Investor magazine, analyzed in Erb et al. (1996)) play a clearly subordinate role. Bhatia (2002) gives an overview of the ratings methodology and points out several explanations why ratings may fail to foresee crises. Hampered by the statistically small number of sovereign crises and their ever changing nature, the quality and timeliness of sovereign ratings as proxy for sovereign default risk is difficult to assess. Reinhart (2002) identifies ratings as a poor predictor of currency crises (which, in turn, are often followed by sovereign debt crises).

While bond spreads and ratings do not always cohere (Sy (2002)), rating actions are found to be reasonable predictors of sovereign distress (Sy (2004)). Although market participants proved sceptical about the quality of rating assessments a decade ago (Group of Ten (1996), p. 30), this situation may have improved as the market expanded.

The determinants of sovereign ratings closely resemble those of bond spreads (Bissoondoyal-Bheenick (2005), Cantor and Packer (1996), Hu et al. (2002), Larrain et al. (1997), McNamara and Vaaler (2000)). As broadly agreed upon in these studies, sovereign credit ratings can largely be explained by indicators of external debt, fiscal and external balances, inflation, real growth and per capita income. Political variables appear to play a minor role in determining ratings (Haque et al. (1998), Block et al. (2003)). This may arise from difficulties in measuring the effects statistically although politics are undoubtedly of pivotal importance in emerging markets. The spillover effects of negative rating changes are analyzed by Gande and Parsley (2005).

While the information contained in ratings is found to be made up by fundamental indicators, the rating assessment itself is believed to contain additional information (Cantor and Packer (1996)). Event studies assert that bond spreads, as well as other financial markets, react significantly (but maybe not homogenously) to rating announcements (Andritzky et al. (2005), Norden and Weber (2004), Reisen and von Maltzan (1999)).

2.4.4 Determinants of Spreads

Secondary market prices of sovereign bonds offer a continuous measure of sovereign default risk. Since the emergence of sovereign bonds on the international capital markets, a burgeoning literature has researched the determinants of these credit spreads.

Even before the first Brady deals, early contributors analyzed the explanatory factors of risk premia in foreign loans issued by less developed countries (Edwards (1986), Boehmer and Megginson (1990)). By comparing the determinants of LIBOR spreads of loans and offering spreads of bonds, Edwards (1986) shows that similar fundamental drivers are at work. This provides the first evidence of the parallels between the determinants of crises and ratings on the one hand, and spreads on the other. Many authors have confirmed this stance (see Tab. 2.4). The close relationship between fundamentals and credit ratings has proven the latter to be a good substitute for the former (Cantor and Packer (1996)), causing some empirical studies to rely solely on ratings as a proxy for all fundamentals (e.g., Kamin and von Kleist (1999)).

Resorting to spreads as the dependent variable allows for the estimation of the "cost" of a change in fundamentals or ratings, e.g. the average increase of spreads in response to a rating change (Sy (2004)). In the same way as past defaults are identified to increase spreads (Cantor and Packer (1996), Eichengreen and Mody (1998b)), some evidence hints at differences in spreads due to regional origin (Kamin and von Kleist (1999)) or instrument characteristics (e.g., Eichengreen and Mody (2000b,c)).

Fundamental factors alone, however, reveal only one part of the picture. Calvo et al. (1993) refer to these as country-specific "pull" factors, which operate alongside global "push" factors. The importance of general "market sentiments" was recognized by a number of commentators (Eichengreen and Mody (1998b), Kamin and von Kleist (1999)). One string of the subsequent literature focused on the impact of world interest rates. While theoretical arguments suggest a straight-forward relationship, the empirical literature could not always provide unanimous evidence. Looking at capital flows to developing countries, the search-for-yield hypothesis proclaims that lower world interest rates have a positive impact on the demand for emerging market investments. Empirically, this is well supported (Dooley et al. (1994), Eichengreen and Mody (1998a)) but poses a difficult policy dilemma for the monetary authorities in industrialized countries, as the findings lend support to capital-flow induced boom-and-bust cycles in developing countries. Besides diluting the value of economic adjustment programs, the finding questioned the advisability of capital-account liberalization (Kenen (1998)). This discussion toned down as the empirical literature remained inconclusive about the importance of world interest rates on emerging market bond spreads (in contrast to capital flows), which are considered an indicator of market access. In theory, lower world interest rates are argued to lower risk spreads (as lower borrowing costs to the debtor decrease the likelihood of default), and increased risk appetite fuels the demand for riskier investments. Inconclusive empirical evidence in this area highlights the difficulties of spread analyses. Early studies found little evidence for the effect of world interest rates (Cline and Barnes (1997), Kamin and von Kleist (1999)), while more sophisticated studies of secondary spreads are generally more in line with the aforementioned theory (Arora and Cerisola (2001), Ferrucci (2003), Uribe and Yue (2003)).

Table 2.4. Determinants of emerging market bond spreads

	Primary market bond spread levels			Secondary market country spread levels 1/				Changes in spreads 2/
	Edwards (1986)	Eichengreen and Mody (1998b)	Min et al. (2003)	Cantor and Packer (1996)	Arora and Cerisola (2001)	Sy (2002)	Ferrucci (2003)	Westphalen (2001a)
No. of countries	13	37	11	35	11	17	23	26
Start	1976	1991	1991	1995 /3	1994	1994	1991	1995
End	1980	1995	1999		1999	2001	2003	2001
Bond features								
Issue size	–							
Maturity or duration		+				–		
Solvency indicators								
Debt load 4/		+	+	+	+		+	
Rating 5/		–				–		
Liquidity indicators								
Debt service ratio 6/		+	+		+			+
Reserves 7/			–		–			
Pull factors								
GDP growth		–						
Inflation			+					
Fiscal balance					–		–	
Default history 8/		+		+				
Push factors								
Global interest rate		–	+		+	+	+	
Adj. R^2	0.44	n/a	0.65	0.86	0.37-0.82	0.84	0.40	0.16

Sources: author's compilation. Positive and negative signs indicate significant regression coefficients. 1/ Monthly frequency of composite spread index, mostly EMBI country spreads. 2/ Monthly spread changes of single issues on the secondary market. 3/ Spread levels on 29 September 1995. 4/ Mostly external debt as ratio of GDP, GNP or exports. 5/ Translated into linear scale with higher values indicating higher credit standing. Partly representing only the residual component from a regression of ratings on fundamentals. 6/ Interest and principal payments as ratio of exports or GDP. 7/ As ratio of imports or total external debt. 8/ Mostly dummy variable for past defaults.

While this question has important policy implications, a holistic picture of spread determinants of sovereign bonds is still missing, at least at a level of accuracy useful for bond pricing.[45] A first shortcoming lies in the fact that most fundamental determinants are available with limited frequency only, such as monthly or quarterly reported indicators. This situation is mitigated by the emergence of continuous proxies of fundamentals, such as local stock market indices or exchange rates. The development of more sophisticated proxies of push factors, such as global liquidity measures and risk aversion indices, has lent significantly to the improvement of explanatory power in empirical models. Global market integration is expected to increase the importance of these drivers (McGuire and Schrijvers (2003)). A second, often underestimated challenge, is the efficient and complete use of information contained in financial data. Although data limitations remain a severe constraint, the econometric exploitation of vast amounts of financial time series is likely to gain importance as financial markets in emerging countries grow and mature. While studies relying on composite indices (that mix different bond seniorities and durations) are valid starting points, a better fit of single bond issues can only be provided by term structure models (such as Duffie et al. (2003)). These models work with unobservable stochastic variables as drivers of bond spreads. Unfortunately, approaches to link these variables to macroeconomic fundamentals are still in their infancy (Diebold et al. (2005)). Ang and Piazzesi (2003) present a first and promising approach by means of a vector autoregressive model. They find that macroeconomic factors can explain up to 85% of variation in bond yields, but mostly at the short end and middle of the yield curve. Unobservable factors remain the major drivers of movements at the long end.

2.5 Concluding Remarks

The historical perspective suggests that the recent expansion of emerging market lending, and the recent crises, are not per se exceptional developments. On the contrary, the recent discussion on an overhaul of the current financial architecture has brought up suggestions that resemble previous arrangements. Creditor coordination, for instance, was addressed by bondholder committees over a century ago. State

---

[45] This induced studies that found systematic persistencies in the pricing of bonds, usable for profitable trading strategies. See Berardi et al. (2004), or Jostova (2006).

contingent debt service has been utilized earlier by means of equity instruments and is reinvigorated by indexed bonds (such as GDP-linked warrants) today. With the development of new markets, such as domestic debt markets in emerging countries, one-off initiatives for solving debt problems, comparable to the Brady plan in the 1990s, will become possible again.

The new stance towards private sector involvement has alerted many sovereign bond investors. This concern is at least partly justifiable. On the one hand, the inclusion of sovereign bonds in crisis resolution may indeed create some costs to investors when a crisis cannot be averted. An event study by McBrady and Seasholes (2004) estimates the effect of comparable treatment of private bond holders, first applied in Pakistan 1999, to increase spreads by 25 to 95 basis points, at least in the short-run. This is in contrast to the early 1990s, where sovereign bonds could evade the restructuring efforts. On the other hand, the burden sharing principle may increase the IMF's capacity to prevent and contain crises. Furthermore, IMF involvement may enhance the efficiency of bond restructurings (e.g., by facilitating creditor coordination), and therefore present a mutual benefit for creditors and debtors. This policy change, supported by the G-8 leaders and withstanding court rulings of major financial centers, creates the opportunity for sovereign debtors in distress to restructure and consolidate their bonded debt. Some might be able to partly convert bond obligations into local currency or, at least, add collective action clauses.

The scientific literature has tried to reproduce the forces at work in theoretical models and detect common grounds in empirical studies. While there is, broadly speaking, a general consensus on the determinants of crises, ratings, and composite bond spreads, it remains difficult to determine what drives single spread movements and to explain the full extent of the term structure. Similarly, forecasting threshold values for crises, and making predictions about the sequence and timing of distress events unfolding, currently seems beyond the scope of scientific cognizance.

A number of specific issues remain to be solved. First, political factors deserve larger attention. An exception is the discussion on moral hazard induced by IMF lending, a discussion which has been set off track through the introduction of the "Mussa argument" by Jeanne and Zettelmeyer (2004). A more comprehensive picture on links between politics and sovereign default risk, however, is missing. Although market experts are well aware of the impact of domestic and international politics, especially with regard to emerging market financial instruments, the empirical literature on the political economy of sovereign debt is relatively new (e.g., Block and Vaaler (2004)). The discussion on sovereign restructuring mechanisms and the legal features of bond contracts (such as collective action clauses) has already promoted empirical work in this area (e.g., Eichengreen and Mody (2000b,c)). By presenting a thorough analysis of the structure of recent sovereign debt workouts, Chap. 3 sheds more light on this area. The findings might be of use for the development of models that are coherent with the stylized facts.

Second, a more thorough knowledge of feedback effects is necessary to understand bond spreads, themselves being a far forward-looking measure of credit risk. Eichengreen and Mody (1998b) stress supply versus demand effects, showing that the likelihood of a bond issue declines with higher U.S. Treasury yields. Similar effects are at play during crises, which reduce a country's ability to issue debt. Sy (2004) shows that capital market access is reduced by one half when spreads rise above 1,000 basis points. Detragiache and Spilimbergo (2001) argue that the accumulation of short-term debt is endogenous during the run-up to a crises as countries find it more difficult to borrow on longer maturities. The study in hand contributes in this area by providing alternatives to assuming an exogenously given and deterministic recovery value, a shortcoming of most current studies which look at spreads only. Determining the drivers of the expected recovery value, i.e. making the recovery value endogenous, can help to enhance asset pricing of bonds and credit derivatives (see Chaps. 4 to 6).

3

Sovereign Restructuring

The restructuring of sovereign bonds is a new and urgent issue for the financial community. The hope that default will become an unheard-of word on the international market for sovereign bonds did not come true in the last decade. The official sector's call for private sector involvement (PSI) marked an end to unbounded rescue efforts and forced private creditors to contribute to the resolution of debt crises. Creditors, oftentimes rewarded for investing in emerging countries through considerable risk premia, became concerned that they would have to face significant forfeits in upcoming debt swaps or restructurings. But are these restructurings worth the fear investors have about them? This question is addressed in the following three sections. A review of the literature in Sect. 3.1 explains the challenge of workout mechanisms and describes the most recent proposals discussed in politics and science. The subsequent Sects. 3.2 and 3.3 generalize the circumstances of debt restructurings and compare the terms of recent transactions.

Section 3.4 sheds light on the implications on investment returns and addresses the question of how to evaluate a restructuring offer. Applying a 10% flat discount rate and comparing the resulting net present value (NPV) is generally too simple. The more appropriate way to evaluate restructurings ex post is the comparison of holding period returns of different periods throughout a restructuring. This takes into account that risk spreads decline from their peaks after a successful restructuring.[1] The results show that the most significant losses are realized during the early stages of the crisis, whereas a restructuring event itself turns out to be mutually beneficial for the debtor and the

---

[1] The approach is related to the concept of the trading price recovery which regards the post-default market price of securities as recovery value. See Renault and Scaillet (2004), p. 2920.

creditor. In the run-up of a crisis, market valuations seem to be too pessimistic and trading liquidity becomes thin (indicating loss of market access) whereas, after a restructuring, trading of the restructured bonds is regaining liquidity.

The insights gained in this analysis feed back into pricing models. To come up with a suitable model of sovereign default risk, the final part of this chapter addresses the question, which recovery assumption is reasonable for a sovereign debtor? While there is considerable expertise from realized corporate defaults, making it easy to pick and justify a certain recovery expectation for valuation purposes, this is not the case for sovereign bonds. Frequently, bond market analyses rely on the common recovery of market value (RMV) assumption, probably due to its simplicity.[2] The alternative concept of recovery as a fraction of face value (RFV), however, must not be disregarded. When taking a look at the recent restructurings, it appears reasonable to take a combination of both concepts into account, depending on the circumstances. A restructuring with the sole objective of prolonging maturities and providing short-term relief resembles an RMV-like recovery. Restructurings by a defaulted debtor involving haircuts seem more closely related to the RFV concept.

3.1 Literature Review

The literature on aspects of sovereign debt restructuring is expanding very quickly as the discussion on the appropriate restructuring mechanism continues. The contributions to this topic are manifold and involve economic, legislative, and political aspects. The review is divided into two parts. In the first part, the development of ideas for bankruptcy procedures of sovereigns is reviewed. The second part looks at the few studies conducted on recent restructurings.

3.1.1 Sovereign Bankruptcy Procedures

Dealing with sovereign bankruptcy goes far back in history, as Chap. 2 has shown. Ever since that time, the sovereign workout process was perceived to bear significant inefficiencies. Most of these inefficiencies consist in coordination problems and distorted incentives, additional to the large degree of uncertainty inherent with sovereign bankruptcies. The literature has therefore centered on proposals how to make a workout as efficient as possible.

---

[2] See Duffie and Singleton (1999), pp. 700ff.

Thereby, well designed workout processes need to comply with the following criteria. Firstly, any workout process has to avoid debtor and creditor moral hazard and ensure a timely bankruptcy declaration. Making default very costly to the sovereign can avoid the incentive of over-borrowing and opportunistic default.[3] However, this may create little incentive ex ante against over-borrowing because taking on debt by the incumbent government creates an instant benefit while the burden of repayment may be inherited by another administration in the future.[4] During a crisis, the high cost of default can also give policymakers reasons to delay the initiation of restructurings beyond a desirable point. When a restructuring is clearly inevitable, this creates only more hardship for the domestic economy, and possibly reduces the sovereign's debt servicing capacity in the future. Secondly, the workout should be comprehensive. The principle of burden sharing requires that all creditors contribute to the crisis resolution effort. At the same time, creditors of one class of debt instruments (for example, pari passu ranking bonds) must be treated equally. This prevents, for instance, domestic creditors or banks to receive preferred treatment. Thirdly, the workout process should prevent self-fulfilling creditor runs.[5] One is the "rush to the exists" upon first signs of a crisis, creating large capital outflows and possibly causing a liquidity crunch. Another kind of run is the "rush to the courthouse" when creditors try to move quickly to seize the debtor's assets, undermining comprehensive renegotiation efforts.[6] Another creditor coordination problem arises from informational asymmetries and free-rider behavior, like investor holdouts. This term refers to a single creditor or a minority group of creditors who are free-riding on a restructuring agreement by staying away from the exchange while enforcing their payments on the original bonds later, often at the expense of those creditors who already accepted the exchange offer. Information asymmetries arise from missing incentives on either the debtor and creditor side to reveal the "true" payment capacity or minimum conditions for accepting a restructuring deal, respectively.[7] Lastly, the process should be concluded within a reasonable time frame.

A large number of suggestions exist on how to draft a workout process which fulfills these criteria. Rogoff and Zettelmeyer (2002) offer an excellent overview. In the postwar era, the discussion started with the

---

[3] See Dooley and Verma (2001).

[4] See Alesina and Tabellini (1990) and Corsetti and Roubini (1997).

[5] See, among others, Chang and Velasco (2000), Cole and Kehoe (2000), Detragiache and Spilimbergo (2001).

[6] See Miller and Zhang (2000).

[7] See Haldane et al. (2005).

debt problems of many less developed countries (LDC) in the 1970s and 1980s and continues today, while emphasis of the proposals shifted from addressing one to addressing another of the above criteria over time. Since the beginning of the discussion, proposals of an independent arbitrator with sufficient power vis-à-vis debtors and creditors kept coming up again and again, but remained utopia up to now. In 1979, the Group of 77 developing countries first proposed such a centralized body, the "International Debt Commission", which would follow the style of Paris Club negotiations. Oechsli (1981) constitutes the first comprehensive proposal on a supranational bankruptcy procedure derived from Chapter 11 of the U.S. bankruptcy code.

By the mid-1980s, the large debt burden of many developing countries was believed to curtail their growth due to a "debt overhang" problem. In terms of a debt Laffer curve, this suggests that the disincentives of debt have grown so large that debt relief would actually increase the default risk-adjusted value of LDC debt (Krugman (1988), Dooley (1986)). This created a wave of market-based debt reduction schemes, for instance, by converting external debt in domestic equity at a discount. While some criticized that this mainly benefited creditors, this approach of uncoordinated, voluntary debt reduction proved inefficient due to the large degree of free-riding (Bulow and Rogoff (1991), Dooley (1988), Helpman (1989)). Despite several authors drafting new statutory workout proposals to mitigate this problem (Barnett et al. (1984), Cohen (1989a,b), Raffer (1990)), the Brady plan put an end to this discussion in 1989. Although the plan presented only a one-off initiative, it provided the background for some much needed change to the international financial architecture. The most important one was the new stance towards IMF lending into arrears, which strengthened the IMF's role in crises and workout processes. In turn, this lowered the bar for (unilaterally announced) debt moratoriums or "standstills", giving the debtor more breathing space and forcing the creditors to the negotiation table.

Defying these improvements, the reemerging crises in the 1990s motivated a new string of proposals, now stressing moral hazard issues and the avoidance of investor runs. This discussion has abated today, but initially centered on two main alternatives for debt workouts. The centralized approach envisages the creation of a central bankruptcy court for sovereigns, applying generally accepted principles of a supranational bankruptcy code. While a neutral and completely independent body might be desirable, the most likely way for implementation would consist in an amendment of the IMF Articles of Agreement. The latest proposal heading this direction was made by the IMF's First Deputy Managing Director Anne Krueger, suggesting the "Sovereign Debt Restructuring Mechanism (SDRM)" (Krueger (2001)) with the following main features:

General voting rule. Decisions by a qualified majority of creditors ("supermajority") becomes binding for all claims.

"Dispute Resolution Forum (DRF)." An independent committee is appointed which serves for administrative functions, dispute settlement, and enforcement actions. The DRF assures the transparency of the process and the disclosure of key data about the debtor.

Temporary stay. Upon a motion of the defaulted debtor, a supermajority decides about a limitation on creditor enforcement.

Priority finance. By majority voting of the creditors, a preferred creditor status is approved for interim emergency loans ("fresh money") to overcome a liquidity squeeze.

Constraints on the debtor. This extends the idea of IMF conditionality in order to avoid free-riding, for example by imposing capital controls to stop capital flight.

This draft of a centralized workout body shows parallels to many other proposals made in the present and the past. However, emphasis is placed on the avoidance of large bailouts as they exhaust the IMF's financial capacities and potentially create moral hazard. In the Krueger (2001) proposal, an immediate, IMF-endorsed moratorium (before any IMF support is granted) is believed to eliminate panics and reduce the need for large bailouts.[8] The implementation of such a statute, however, appears politically infeasible as a wide-reaching consensus on its necessity is lacking. Prominent commentators and market sources

---

[8] Others, for instance Lerrick and Meltzer (2001), believe that the collapse of secondary debt market prices mainly causes panic and contagion. They suggest addressing this problem directly, by introducing an "offical floor of support" with unconditional IMF lending to keep secondary market bond prices at levels which are believed to reflect a sustainable debt burden during a restructuring. The price collapse of bond prices, however, is not the technical reason behind a liquidity squeeze. What matters are other channels of capital outflows and the prohibitive marginal cost of new lending (which is better mitigated by IMF emergency lending than by an IMF intervention on bond markets). However, it is believed that this could help to avoid irrationally low secondary market prices attracting vulture funds and other speculative interests which may later disrupt an orderly workout.

criticize the proposal as regulatory overkill that could hamper future capital flows into these markets.[9]

As opposed to the centralized approach, the decentralized approach stresses a market based solution. This approach mainly relies on the introduction of appropriate clauses in bond contracts. While the idea was much promoted by the former U.S. Treasury Undersecretary John Taylor (Taylor (2002)), the potential role of majority clauses in bond contracts was originally emphasized by academic authors, such as Eichengreen and Portes (1995). The clauses, mostly referred to as collective action clauses (CAC), provide for the following features:

Majority action. Majority voting permits the amendment of bond terms in a restructuring by the consent of a supermajority of at least 75%.

Engagement. Engagement clauses provide precepts for the appointment of a representative bondholder committee which may set further internal rules.

Initiation. This clause determines voting requirements for the acceleration in a continuing event of default or the annulment of acceleration when the default is cured. Most contracts use a 25% voting threshold for acceleration and a 50% to two-thirds majority for de-acceleration.

Transparency. The issuer is required to follow certain disclosure requirements and provide key data.

Collective action clauses in bond contracts are a common feature under a number of legislations, such as British law. A bond issue by Mexico in 2003 has shown that including CACs in bond contracts under New York law is legally feasible.[10] Others have followed that route. Today, more than 90% of all sovereign bond issues under New York law include CACs which are now considered market standard.[11] Besides New York, it seems legally viable to include CACs under other legislations as well, such as Germany.[12] Certainly, the inclusion of CACs is an endogenous variable and cannot be enforced. Nevertheless, it seems that CACs are getting more popular and will become a global market standard. Initially, the driver behind this trend seemed to be the "threat" of a

---

[9] See, for instance, Shleifer (2003), Eaton (2002), and Group of Ten (1996).

[10] New York is the dominant jurisdiction for the sovereign bond primary market, holding a market share of two-thirds by market value in 2004.

[11] See International Monetary Fund (2005).

[12] See Becker et al. (2003).

statutory workout mechanism as alternative solution.[13] It will, however, need several years before all existing bonds will be replaced by contracts including CACs. Bratton and Gulati (2004) calculate a penetration rate of CACs (given all new issues include the clauses) of 80% by 2010, or maybe higher if sovereigns decide to launch bond exchanges before maturity.

After discussing the dominant proposals for sovereign workouts, the following turns to the academic contributions on analyzing these approaches. Thereby, the literature on the best approach to sovereign restructuring is broadly divided into two strands. The first builds on theoretical arguments and models.[14] Among those, Kletzer (2003) and Haldane et al. (2005) show how unanimous voting leads to protracted negotiations and strategic rent seeking behavior, resulting in inefficient outcomes. Under ideal conditions, CAC-induced majority voting results in the first-best solution to the problem, rendering any statutory restructuring redundant. To avoid failures of creditor coordination across bond issues and jurisdictions, Kletzer (2003) proposes the use of the aggregation clause. This clause requires a majority consent by all pari passu ranking bondholders (as opposed to majority votes of each single bond issue).[15] If, however, incomplete information about the investors' valuation of the offer prevails, uncertainty will trigger strategic behavior. In this case, Haldane et al. (2005) suggests an intermediator (like the IMF) to observe the true valuation in a debt exchange, balance the interests, and achieve a social optimum. Weinschelbaum and Wynne (2005) highlights the aspect of free riding among creditors of different jurisdictions. They argue that CACs are irrelevant under the current condition in which bond contracts and jurisdictions widely differ and the presence of IMF bailouts distorts investor incentives.

The second stream of the literature researches empirically the effect of CACs on bond market spreads. Initial studies failed to find any significant effect when comparing bond spreads with and without collective action clauses on the primary and secondary markets (Becker et al. (2003), Dixon and Wall (2000), Gugiatti and Richards (2003)). The approach by Eichengreen and Mody (2000b, c) and Eichengreen,

---

[13] After the successful placement of CAC-bonds in New York in end of 2003, Guillermo Ortíz, governor of the Bank of Mexico, is quoted: "We were worried because it would increase our financing costs. The truth is we did it because it was a way to get rid of the SDRM." Galloway (2003), p. 24.

[14] See, besides the contributions described in the following, Thomas (2004), Eichengreen, Kletzer and Mody (2003), and Ghosal and Miller (2003).

[15] Similar clauses have recently been implemented after restructurings in Uruguay, Argentina, and the Dominican Republic.

Kletzer and Mody (2003) recognizes the endogeneity problem since borrowers can choose the contract design (which might constitute a signal), and both lenders and borrowers decide on whether to approach the markets at all. The result suggest that CACs contribute towards lower spreads for debtors with good credit standing while increasing spreads for weak debtors.

3.1.2 Analyzing Past Workouts

Only a very limited number of studies, though, analyze recent sovereign restructurings. Thereby, most contributions focus on the question of how the official sector should respond to debt crises, but give little hindsight what investors can expect from sovereign bond restructurings. In a paper about investor bail-in (the dominant term at that time), Eichengreen and Rühl (2000) offer a review of the resolution of debt crises in Pakistan, Ecuador, Romania, and Ukraine. The paper by Sturzenegger (2002), prepared for the World Bank, is the most comprehensive review of sovereign restructurings but does not yet include the cases of Argentina and Uruguay.[16] A more systematic comparison is found in Palacios (2004). The Russian and Argentine debt swaps are described in the paper by Aizenman et al. (2005), providing a model of debt swaps.[17] Further details about these restructurings are included in a number of other publications, such as Lipworth and Nystedt (2001). Noteworthy are the policy-considerations by the IMF.[18]

In the area of sovereign bond investments, literature focusing on portfolio management is rare. A comprehensive study of how much investors have earned in emerging market investments is found in Klingen et al. (2004). Sturzenegger and Zettelmeyer (2005) analyze the haircut implied by recent restructurings by using the post-restructuring exit yield, i.e. the yield at the first trading day after the restructuring, to value the pre-restructuring original bonds. Comparing them to the market value of the restructured instruments can make successful restructurings appear especially painful. If the restructuring reinstalls long-term debt sustainability and exit spreads come out very low as a result, the yield-implied value of the (more generous) pre-restructuring bond terms would suggest a larger price reduction than investors actually suffered.

---

[16] A redrafted version is available as Chuhan and Sturzenegger (2003).

[17] The description of the debt swaps draws on Kharas et al. (2001) (Russia), as well as Perry and Serven (2003), Mussa (2002), and De la Torre et al. (2002) (Argentina).

[18] See International Monetary Fund (2001, 2002, 2003)

The changing nature of the young mechanisms of sovereign bond restructurings awaits more scientific analyses. The following presents one attempt, trying to identify common patterns in the restructuring mechanics which allow the reshaping of valuation models for sovereign bonds.

3.2 Crisis Resolution in a Nutshell

While each approach to debt crises is as unique as the underlying economic and political problems, it is nevertheless helpful to establish a structure for classifying debt crises in order to find an appropriate solution. Figure 3.1 presents such an overview compiled on the base of Manasse and Roubini (2005), Rieffel (2003), Roubini and Setser (2004a), and others.

3.2.1 Liquidity and Solvency Crises

Financial crises are often distinguished between liquidity and solvency crises, a concept vitalized in Corsetti et al. (2005). Even though this categorization is easily done for corporate entities based on their balance sheets, the solvency concept is blurred for sovereign entities. Definitions on the distinction therefore start from the symptoms, not their causes.

Liquidity crises are debt problems marked by the inability to roll over, i.e. to refinance short-term liabilities. Liquidity crises can seamlessly spill over into a solvency crisis. In fact, while not every unsustainable debt level immediately leads to insolvency, any solvency crisis is triggered by some kind of refinancing shortage. Although causes and indicators of both types of crises therefore look closely related, the severity and the medium- to long-term perspective of a foreseeable or current refinancing problem determines the appropriate measures to be taken. A momentary refinancing problem—perhaps caused by a slump in demand for emerging market debt as a result of contagion or unexpected policy moves by the Fed—can be overcome in a concerted action such as a stand-by assistance from the IMF, a commitment from commercial lenders to roll over loans, a rescheduling at the Paris Club under "Classic Terms" (without any reduction of debt obligations), and a voluntary swap of short-term debt securities into longer maturities.

Situations stemming from refinancing problems can be identified by a number of indicators. While the overall debt load is not excessive, the debt servicing profile might be front loaded, meaning that considerable payments have to be made in the short run. This was the case,

Fig. 3.1. Sovereign debt crises and solutions

Liquidity crisis	Solvency crisis
Causes: • Considerable rollover risk • Confidence crisis • Contagion • Rising benchmark rates • Low demand for emerging market debt	Causes: • Economic depression or transformation • Costly domestic crises, such as banking crises • Political crisis • Currency crisis and currency mismatch • Unsustainable debt level
Indicators: • Front loaded public debt service profile • Inverse term structure of sovereign bonds • Low international reserves in relation to short term obligations • High reliance on capital account surpluses • Weakening exchange rate	Indicators: • High external debt-to-GDP ratio • Inverse or sharply hump-shaped term structure • High inflation • Low level of government revenues and sustained fiscal deficit • Current account imbalance • Exchange rate misalignment
Examples: Mexico (1995), Ukraine (1999), Brazil (1999)	Examples: Ecuador (1999), Pakistan (1999), Argentina (2001)

Debt swap	Soft restructuring	Hard restructuring
Objectives: • Maturity extension • Short-term cash flow relief	Objectives: • Short-term cash flow relief • Medium-term debt sustainability • Economic or political benefits	Objectives: • Long-term debt sustainability, debt relief • Economic or political turnaround • Prerequisite for bi- and multilateral relief or assistance
Indication: • Debt service load caused by single bond • Definable bondholder group	Indication: • IFI support • Diverse bondholder universe • Collective action and aggregation clauses	Indication: • Ultimate measure, market access lost, default inevitable • Deep economic or political crisis
Characteristics: • Voluntary exchange, i.e. tender • Discriminatory eligibility • Instrument upgrade, sweeteners	Characteristics: • Market sounding, informal negotiations • IFI involvement • Menu approach, sweeteners • Take-it-or-leave-it offer, latent default threat • Exit consents by majority vote	Characteristics: • Default • Face value, coupon, and PDI reduction • Take-it-or-leave-it offer • Exit consents • Hold out threat and litigation
Key Success Factors: • Participation rate • Moderate yield and face value increase • No selective default rating	Key Success Factors: • Low ratio of holdouts • Reduction in spreads and market access • Improvement of rating	Key Success Factors: • Debt relief • Low level of litigation • Reduction in spreads and market access • Improvement of rating • Timely implementation
Examples: • Ukraine and Regent Pacific Group (1999) • Argentina (July 2001) • Moldova buybacks (2001) • Turkey (2001), Lebanon (2004)	Examples: • Pakistan (1999), Ukraine (2000) • Uruguay (2003) • Dominican Republic (2005)	Examples: • Ecuador (1999) • Russia (1998-2000) • Argentina (2001-2005)

Source: Author's illustration.

for instance, in the "Tesobono Crisis" in Mexico. Manasse and Roubini (2005) find in their empirical assessment of crisis indicators that pure liquidity crises occur at total external debt-to-GDP ratios of less than 49.7%, while typically the short-term debt-over-reserves ratio is larger than 1.3.[19] With respect to bond markets, this situation typically occurs when major sovereign bonds are to mature in the near future. If particular bond issues can be identified to cause a potential refinancing problem, financial markets will trade these issues at a discount, often resulting in an inverse term structure of risk premia. Manasse and Roubini (2005) identify that in certain combinations with other indi-

---

[19] Note that these debt measures used by Manasse and Roubini (2005) include both private and public debt.

cators, fixed exchange rates and political uncertainty can be associated with upcoming liquidity problems.

A solvency crisis, in contrast, is characterized by a high debt load. Manasse and Roubini (2005) identify indicator thresholds for insolvency-prone countries, such as an external debt-to-GDP ratio of above 49.7%, public external debt over revenues above 215%, and inflation above 10.5%. With already weak fundamentals, initial refinancing problems often culminate into insolvency upon the arrival of an additional negative shock. This is characteristic for the breakout of most crises and a general attribute of "catastrophic" events. Therefore, an analysis of solvency risk is well advised to include a "360 degree" review of the economy, including the banking system, political stability, monetary policy, and the currency regime.

The determinants just discussed blend differently into the assessment of sovereign risk—such as those by investment banks and asset managers, rating agencies, and the international financial institutions (IFI). However, none of these concepts—neither the clear-cut letter grades from rating agencies nor the diplomatic phrases used in IMF reports—can provide the financial community with a cast iron proof of an imminent liquidity or solvency crisis. As there is a continuum between the three stages "no sovereign crisis", "liquidity crisis", "solvency crisis", the appropriate measures for crisis resolution present a complementary continuum ranging from voluntary debt swaps through harmless soft restructurings to painful hard restructurings. It has to be kept in mind that the measures in the lower part of Fig. 3.1 only present a small section of possible measures used to address a crisis. Since the topic of this study focuses on sovereign bonds, these measures mainly address problems of public debt (in contrast to private sector debt) on the international capital markets (as opposed to loans and domestic bonds, although those debt workouts may look similar).

3.2.2 Debt Swap, Soft and Hard Restructurings

By the end of the 1990s, securitized debt made up a significant share of sovereign indebtedness for many countries, and the traditional lenders of last resort—like the G-8 countries and multilateral agencies—became unwilling and unable to avert major crises. This gave rise to the principle of "burden sharing" and has since legitimated debtor countries to initiate capital market transactions for relief. Therefore, any kind of private sector involvement will take the form of a voluntary debt swap or some kind of comprehensive restructuring.

A **maturity debt swap** intends to exchange short-term liabilities into new instruments which provide short-term cash flow relief.[20] When international capital markets currently do not allow the issuance of new debt, a voluntary debt swap can skim those investors who are interested in prolonging their exposure. From the issuer's point of view, a debt swap should be chosen if a refinancing problem is foreseeable and is caused by a few bonds only. Furthermore, a debt swap is more easily implemented if the eligible issues are held by a definable bondholder group, facilitating creditor coordination.[21] However, it must not be neglected that a debt swap tender comes at some costs as well, and does not contribute to long-term debt relief as new instruments with a longer maturity usually imply a higher yield.[22]

From the investors' point of view, a debt swap offer is of a voluntary nature where bondholders tender their old instruments for new issues.[23] Its impact on the value owed to creditors is usually of small magnitude. The effect on the yield curve depends on the extent of the debt swap, but is generally assumed to be small. Similar to restructurings, most offers include small incentives ("sweeteners") for tendering into the exchange.

What makes a voluntary debt swap offer successful? While the participation of the vast majority of bondholders is not necessary for conducting a debt swap, it is necessary to yield a substantial cash flow relief. However, this is a result of the value offered and therefore presents a

---

[20] Since this study looks at external debt, i.e. debt in foreign currency only, currency debt swaps are of no interest here. Hence, the term "debt swap" refers exclusively on maturity debt swaps. The Russian swap of ruble treasury bills (GKOs) into eurobonds had both features, converting the debt into foreign currency and prolonging maturities. See Kharas et al. (2001).

[21] This was the case in Ukraine, where the Regent Pacific Group was basically the only investor in an $163 million loan placed through ING Barings due in 1999. During the grace period and under the threat of cross-default and acceleration invocation, an agreement between the Ukrainian Ministry of Finance and the investor group was reached. Regent Pacific received a 20% cash payout while the rest was converted into D-mark bonds at a factor of 0.943. See Sturzenegger (2002), p. 36.

[22] The $29 billion swap by Argentina in June 2001 is said to have increased the average interest rate by 1.7% and the average yield by 35 basis points; See Cline (2004) and Sturzenegger (2002). Cunningham et al. (2001) show that the Argentine swap in June 2001 reduced short-term spreads but increased long-maturity spreads while the average spread remained unchanged.

[23] As in the Argentine debt swap in June 2001, competitive or non-competitive offers can be placed. Non-competitive offers will be executed at the cutoff price if it exceeds some threshold published in advance. Competitive offers will not be executed if the offer is below the cutoff price.

trade-off between costs and a high rate of acceptance.[24] While a voluntary debt swap is not an unilateral change in contractual obligations—which is considered a technical default by rating agencies—there is, however, a residual risk of receiving a selective default rating. To prevent this, rating agencies require that the offer is public, does not violate contractual commitments, and therefore treats foreign and local investors equally. Attempts that do not honor these rules do not present a sustainable solution since investors may initiate legal action and the international financial institutions might withdraw their support.[25]

If successful and timely implementation appears possible, a debt swap is always a first measure to undertake in order to prevent a refinancing problem.[26] However, this might not always help if a solvency problem persists and economic or political circumstances remain harsh. If a crisis has already advanced too far, a debt swap might rather bind valuable resources and contribute to political inflexibility. The same applies to currency swaps which convert local currency debt earning high local interests into lower yielding foreign currency debt. This intended interest rate arbitrage strategy is dangerous because it increases the expected rate of currency devaluation (and the likelihood of a speculative attack on a pegged rate). This can lead to an even faster meltdown of the crisis.[27] In these cases, the conditions of the debt exchange are unlikely to fulfill the two main characteristics of a debt swap anyway: voluntary participation and selective eligibility.

From an investor's point of view, a debt swap tender must not be expected to heavily impact the valuation of the claims (neither if participating in, nor if abstaining from the offer), although there might be

---

[24] In terms of participation, the Argentine debt swap in June 2001 was a success, exchanging $32.8 billion of face value, a 45% portion of the eligible claims. The downside of this success consisted in its high costs: while debt service obligations until 2005 fell by about $21 billion, total nominal obligations thereafter rose by stunning $66 billion. See Mussa (2002).

[25] The Argentine "Phase one" debt swap in November 2001, for example, targeted local investors. While successful in its intended purpose, the debt swap was judged as discriminatory which lead to a selective default rating and contributed to the withdrawal of the Fund's support, eventually triggering the default on external debt. See Sturzenegger (2002), pp. 70f.

[26] An example might be the Ecuador swap offer shortly after the country already defaulted on some of its Brady bonds. Investors were not satisfied with the voluntary offer and voted for accelerating the bonds (for which only a 25% quorum is necessary), forcing Ecuador even deeper into default and triggering a comprehensive restructuring in 2000. The Ukrainian Regent Pacific deal poses a related example. See Eichengreen and Rühl (2000).

[27] See Aizenman et al. (2005).

a possible redirection of trading liquidity into the new bonds. For this lack of pricing impact, debt swaps are not considered a credit event in the remainder of this study. The following will therefore concentrate on comprehensive and involuntary restructurings.

In contrast to debt swaps, restructurings are outlined by their comprehensive and involuntary manner. While "comprehensive" does not embrace all securitized claims guaranteed by some governmental entity, this term means that all issues of the same class and of the same legal standing (as indicated by a pari passu clause) take part in the restructuring. And while "involuntary" does not mean that investors do not have a choice, it refers to a take-it-or-leave-it offer in contrast to a tender offer.[28]

Preventive restructurings or restructurings on the edge of default can take the form of a **soft restructuring**. Its objective is to overcome short-term refinancing needs and establish a more sustainable perspective in the medium term. Like debt swaps, soft restructurings therefore serve as means to prevent a liquidity crisis and can help to improve medium-term solvency. Since a restructuring is always regarded as an offensive event, sovereigns usually consider it when there are prospective additional benefits. These may be the obtainment of relief or additional assistance, respectively, from bi- and multilateral lenders. Although this should give the country a breathing space, a soft restructuring is not the right measure to simply "buy time" in the hope that economic growth picks up and makes an unsustainable debt burden tolerable. Indications of the merits of a soft restructurings are the establishment or maintenance of cooperation with the international financial institutions. The financial, political, and technical support granted by the bi- und multilateral institutions are often conditional on each other and can be of considerable importance for the fate of a country which is at the verge of a crisis.

Collective action clauses (or majority voting clauses) help to facilitate any restructuring by binding a minority to the exchange and therefore preventing holdouts. An aggregation clause ensures that the majority rule is not only applied to each single issue but to all issues affected by the restructuring. While this may conflict with the equality of treatment (which must be avoided), it offers the chance to consolidate an issuer's bonded debt. The resulting increase in trading liquidity

---

[28] Whether or not a country declares an offer to be "voluntary"—as both Pakistan and Russia did—does not matter. Only by the use of a majority voting rule (which both countries did not invoke) can a minority of investors be bound into an offer. It is impudent, however, to call a democratic rule "involuntary".

benefits all bondholders, while debt swaps and soft restructurings under the unanimity rule expand the universe of bond issues and split up trading liquidity.

The characteristics of restructurings must be seen as a continuum where the soft and the hard restructuring only stand for the purest extremes. The following describes a soft restructuring in an idealized form. In this case, the restructuring offer is preceded by comprehensive market sounding and informal negotiations with investors. Even though the importance of these consultations might be questioned in the light of game theory and political realism, it may fulfill the IMF's requirement for "good faith efforts" and expand the common ground for compromise.[29]

Besides the offering of sweeteners the restructuring offer may include different options to choose from. Such a menu approach has been popular since the Brady plan in order to meet actuarial or tax related needs and other investor preferences.[30] Most debtors predefine some participation threshold for the restructuring offer, although the take-it-or-leave-it character aims at the conversion of all eligible claims. This gives reason to some very typical rhetoric surrounding a restructuring: while creditors or creditor representatives contend to hold out unless the debtor improves the conditions of the restructuring, the debtor establishes a latent threat of defaulting if the restructuring offer is rejected. The only legally binding elements to assign an involuntary take-it-or-leave-it character to a restructuring, however, are so called "exit consents" by majority voting: With their participation in the restructuring, investors are required to agree to changes in the contractual terms of the old bonds. This makes the old bonds less attractive to hold, as the following examples show.[31] Exit consents can be used to invoke collective action clauses by granting the voting rights of the tendered bonds irrevocably to an agent.[32] This way all contractual terms of the original bonds can be changed (or cancelled in exchange for new bonds, as in the Ukrainian case) if a sufficient majority of bondholders agrees to the terms of the restructuring offer. Exit consents can also be applied to claims which do not include collective action clauses. A (simple) majority vote can amend the non-payment terms of the original claims, for example, by deleting the cross-acceleration clause or remov-

---

[29] See International Monetary Fund (2002).

[30] See Buckley (2004).

[31] For a comprehensive assessment from a legal perspective, see Buchheit and Gulati (2000).

[32] Ukraine chose this process for three of their four eurobonds which contained collective action clauses.

ing the covenant to maintain the listing on the stock exchange.[33] The latter reduces the second market value of the original claim which is advantageous for buy-backs by the debtor and discourages investors from holding out. Amending the cross-acceleration clause gives the creditors less leverage and makes it easier for the issuer to default selectively on the old bonds.

If no exit consents are used, the borderline between a voluntary debt swap and a soft restructuring becomes blurred. An offer cannot be attributed to force if the only indicator of involuntariness is of a rhetorical kind. For this reason the case of Pakistan's exchange offer in 1999 represents a borderline case. The Pakistani offer was largely comprehensive—only a small $150 million note was not eligible—and shared other characteristics of a soft restructuring (e.g., the exchange ratio was not determined by a market tender). Furthermore, Pakistan's default threat was absolutely credible. Nevertheless, the offer did not include binding measures against holdouts so that it is hard to argue that it was completely involuntary. The attractive conditions of the restructuring offer—no haircut, instrument upgrade—rather than any kind of force motivated the investors to participate. They acknowledged that the restructuring actually offered additional value. In the end, a vast majority of investors in each issue participated in the restructuring. After closing, Pakistan stayed current on all original bonds, taking no measures against holdouts.

A high participation rate, implying a low ratio of holdouts, is nevertheless an important indicator of a successful restructuring. Furthermore, the restructuring should ideally contribute to a reduction in spreads as the cash flow relief should curb insolvency fears, enabling a debtor to re-access the international capital markets. While typically a country offering a restructuring receives a default-rating, a successful offer should result in an improvement in rating afterwards.

A **hard restructuring** is by all means the ultimate measure to resolve insolvency and comes at high costs for all parties: while the population in the debtor country already suffers from the economic circumstances accompanied with insolvency, default and restructuring comes at high political costs for the country itself and its government, and implies also substantial creditor losses. In this case, the restructuring is a means to make the country current on its debt and restore investor confidence, thereby serving both the debtor's and the creditors' needs.

---

[33] Exit amendments of this kind were first used by Ecuador in 1999.

The objectives of a hard restructuring are obvious. The country's debt has to be resized to a sustainable degree, requiring substantial debt relief. A fresh start has to be made to resolve economic problems and reinstall political order. Private sector involvement requires bondholders to write down part of their claims in order to qualify for debt relief and financial assistance from the bi- and multilateral agencies.

A hard restructuring is usually considered as a last minute measure after other approaches to the problem have failed, market access has been lost, and default on several bonds is inevitable. The restructuring also offers an opportunity to solve an economic or political crisis, for instance, by making adjustments to the currency regime or reforming the politics of fiscal budgeting and tax collection.

In a typical hard restructuring, the issuer already is in arrears with debt service payments and investors already initiated legal action to accelerate their claims. To avoid an investor grab race ("rush to the courtroom"), the defaulting country is well advised to secure its assets against attachment. Drafting the conditions of the restructuring was, up to now, predominantly an unilateral process driven by the debtor country. The spread of collective action clauses and the establishing of bondholder groups with a binding negotiating mandate make it more likely that, in the future, affected investors will be able to influence the process more actively. The notion that the threat of litigation is a major instrument of power in the hands of creditors has not proven true yet.[34]

A characteristic element of any hard restructuring offer is, besides a maturity extension, some reduction in principal value. This may be combined with a reduction in coupon or, as lately introduced by Argentina, a markdown of past due interest (PDI). Holdout investors will not be able to benefit from the improved solvency as free riders, but need to face costly litigation efforts.

Apart from achieving substantial debt relief, reducing spreads and regaining market access, effectively coping with litigation is the most important success factor in hard restructurings. While the attachment of (foreign) assets is still an important lever, litigators frequently target payment flows to the holders of the new bonds, jeopardizing the whole restructuring effort. This can be addressed by measures like exit consents and trust indenture structures (instead of fiscal agents). If the group of holdout investors is small but their litigation efforts prove painful upon the completion of the deal, the sovereign issuer may

---

[34] This view, expressed for instance in Fisch and Gentile (2004), was much influenced by the *Elliott vs. Peru* ruling in 2000, but has ceased ever since.

consider to settle their claims despite the resulting violation of inter-creditor equity.[35] An easy way out is to reopen the exchange under the same conditions at a later time (as happened in Pakistan, Ukraine, and Uruguay).

This framework for resolving sovereign debt problems on the international capital markets recognizes that it is not always possible to draw a clear borderline. The individual characteristics of each crisis and its most appropriate solution are not a discrete choice out of two or three categories. In the future, the whole approach to sovereign restructurings itself is expected to evolve and expand in variety. But the more restructurings we see, the easier it will be to recognize common patterns. While neither all past crises nor all future crises neatly fit this structure, the above framework tries to give some guidelines on the following journey through recent sovereign debt restructurings.

3.3 Evidence From Recent Restructurings

This section compares recent restructurings and relates the commonalities to the general framework of sovereign restructurings. Table 3.1 gives an overview of the main features of the restructurings during the last decade. The table shows the diversity of restructurings following ad hoc rules, but can nevertheless help to distinguish the previously outlined two basic characters of restructurings.

---

[35] A "most favored investor clause" has ruled out this solution in the Argentine restructuring although the robustness of this clause has yet to be tested. The press recently documented efforts by the legal counsel of Argentina to circumvent this restriction. See Diario Ambito Financiero, 5 October 2005.

Table 3.1. Overview of features in recent restructurings of international bonds

	Soft restructurings						Hard restructurings		
Country	Dominica	DomRep	Moldova	Pakistan	Ukraine	Uruguay	Argentina	Ecuador	Russia
Year	2005	2005	2002	1999	2000	2003	2005	1999	2000
Time to closing of exchange									
From credit event (months)	ca. 17 4/	n/a	n/a	n/a	3	n/a	38	11	22
From offer (months)	ca. 8 5/	1	4	1	2	3	2	2	2
Default									
In default during exchange	(Yes)	(No) 7/	(No) 7/	No	Yes	No	Yes	Yes	Yes
Credible threat of default	Yes	Yes	Yes	Yes	n/a	No	n/a	n/a	n/a
Settlement with hold outs	Cont. 6/	Yes	n.a.	Yes	n/a	Yes	No	Yes	(No)
Number of instruments									
Eligible instruments 1/	2	2	1	3	4	18	152	6	6
New instruments	3	2	1	1	2	18	3	2	2
Exchange ratios 2/									
Notional 3/	0.70–1.00	1.00	1.00	1.00–1.057	1.00	0.85–1.08	0.337–1.00	0.65–1.00	0.625–0.67
Past due interest	1.00	n/a	n/a	1.00	1.00 8/	1.00	0.15 9/	1.00	1.00
Change in bond features 2/									
Change in coupon	Yes	No	Yes	Yes	Yes	Partly	Yes	Yes	Yes
Duration extension (years)	7.6	2.0	3.6	3.4	3.0	2.3	8.2	6.7	2.7
Average life extension (years)	9.9	2.8	4.4	4.2	3.9	8.4	16.4	17.0	8.1
Collective action									
CAC included in original bonds	No	No	Yes	Yes	Partly	Partly	Partly	No	Yes
CAC used in exchange	No	No	Yes	No	Yes	Yes	No	No	No
CAC included in new bonds	Yes	Yes	Yes	Yes	Yes	Yes	Yes	No	No
Exit consents used	Yes	Yes	n.a.	No	No	Yes	Yes	Yes	No
Participation rate									
Participation threshold	66%	85%	75%	None	85%	80%	None	85%	86%
Realized participation rate	72%	94%	100%	99%	99%	93%	76%	97%	95%
Reopening of exchange	Yes 5/	No	n/a	Yes	Yes	Yes	No	No	No
Bondholder consultations									
Formal	Yes	No	Yes	No	No	No	No	No	Yes
Informal	No	No	No	No	No	Yes	Yes	Yes	No

Source: Sturzenegger (2002), IMF country reports, prospectuses of offerings, diverse bank publications, author's calculations. n/a: not applicable. 1/ Including Bradies, excluding domestic instruments. 2/ Bradies excluded. 3/ Including cash payouts. 4/ Claims under dispute since January 2003, payments partly halted since July 2003. 5/ Terms of offer have been maintained for quite a while even after the official deadline in June 2004; late participation was possible at least until September 2004. A legal dispute halted the issuance of the new bonds until spring 2005. 6/ Discussions with nonparticipating creditors continue while interest payments at terms of the restructuring are deposited in an escrow account (October 2005). 7/ In arrears on commercial debt, exchange open during the grace period of a missed coupon payment. 8/ Except for Gazprom bonds. 9/ Approximated, based on CSFB.

In the following, the main features of these restructurings are discussed and the resulting "haircut" is assessed on the basis of net present value calculations. The discussion is limited to the restructurings mentioned in Table 3.1 which focuses on securitized debt claims issued on international capital markets. Other cases of private sector involvement, which might have been relevant to local or very specialized international investors, are not considered here.[36] The restructuring and valuation of Brady bonds are not subject of this study and have been analyzed by Buckley (2004) and Claessens and Pennacchi (1996), among others.

3.3.1 Features of Recent Restructurings

Table 3.1 shows that the time between the credit event and the closing of the exchange can differ widely, whereas the actual offering is usually open for four to six weeks. The time which passes until a workout is reached appears to be longer for very disruptive cases. This may stem from the fact that in a full blown crisis the authorities do not assign debt problems the first political priority. In the Argentine case, however, critics argue that the government was reluctant to initiate a restructuring. In all other cases, the length of the time span did not prove a disadvantage to long-term investors since all arrears were compensated in full.

The table also reveals that all restructurings, except for Uruguay, arose from a sovereign default or a credible default threat. While the default is—as outlined in the previous section—the common starting point for a hard restructuring, Ukraine actually defaulted on selective payments in order to demonstrate its need for a debt restructuring. Moldova managed to achieve an agreement on the restructuring during the grace period and is therefore not considered a default.

When dealing with investors not tendering into the exchange, the outcome can be very heterogeneous. In some cases, as in Moldova and for most of the Ukrainian issues, a holdout minority could be bound into an agreement through majority voting clauses.[37] Pakistan remained

---

[36] This results in the exclusion of some bond workouts, such as Russia (GKO and OFZ restructuring in 1998, and MinFin III restructuring in 2000) or Paraguay (default and restructuring of local bonds in 2003), among others. Besides that there are numerous examples of private sector involvement in local debt swaps (Turkey 2001), or bank loans (Ukraine 1998–1999). For those, see Sturzenegger and Zettelmeyer (2005), Cline (2004), and Roubini and Setser (2004b).

[37] There is no confirmed evidence that Ukraine settled with holdout investors in the case of the D-mark eurobond. I thank Rolf J. Koch for providing related legal documentation.

current on all original obligations to avoid litigation. Although preference will be given to the new issues, Uruguay clearly stated from the beginning that debt service on the old claims will be continued. This shows that holdouts did not present a major threat to soft restructurings and were easy to settle; they are obviously a more important issue for hard restructurings. Many suits are pending against Argentina, but up to now no plaintiff has gained a clear victory. While there is no information on holdouts in the Russian restructuring, Ecuador has backed down holdout threats by either settling the accelerated claims or continuing the debt service. While this applied only to a very small fraction of the total debt and did not receive much media attention, vulture investors are well aware of loopholes like this. It is needless to say that full compensation (as Ecuador chose) creates incentives for free riding and is not conducive for avoiding holdouts in future restructurings.

With regard to the number of instruments eligible for exchange and the number of restructured instruments, it is obvious that a consolidation was desired in all cases. On the one hand, this is advantageous to investors as trading liquidity becomes more concentrated. On the other hand, it means that bonds with initially very different contractual features—especially with regard to coupon and currency denomination—might be treated similarly in the restructuring. Uruguay has presented an approach which leaves the choice to the bondholders: investors were able to choose between the option to exchange their bonds into new issues with almost similar features but longer maturity ("maturity extension option"), and the option to exchange them into larger and more liquid bonds ("benchmark bond option"). Almost two-thirds of bondholders chose the benchmark bond option.[38]

As highlighted before, hard restructurings are characterized by write-downs in the nominal value. The haircut in the Argentine exchange marks an extraordinary milestone. By not honoring past-due interest in full, Argentina set a precedent which might be persuasive for future restructurings.[39] In soft restructurings, reductions in par value are avoided. But recent cases might question this borderline as well.[40]

---

[38] Since all benchmark bonds were denominated in US dollars, however, only 22% of the issues denominated in another G7-currency were exchanged into the benchmark bonds.

[39] Argentina, which defaulted in December 2001, was offering cash repayment of past-due interest until 31-Dec-2001 and dated back the restructured bonds as of 31-Dec-2003, effectively repudiating two years of debt service.

[40] In Dominica, investors had the choice between three instruments with the same coupon but different maturities: 10, 20, and 30 years. If an investor prefers the

Other bond features are also subject to modifications. In terms of coupons, restructurings involve all kinds of changes. Only Uruguay's "maturity extension option" tried to maintain the existing coupon. Any restructuring has pursued to expand the bond duration and average life, as is shown in Table 3.2.[41] It is striking that soft restructurings implied only a moderate extension of duration and average life, while hard restructurings seek an almost twice as high duration extension with extremely back-loaded principal repayments. Hard restructurings normally exploit the full maturity spectrum available on sovereign bond markets, which used to be 30 years. With European sovereign bonds now reaching maturities of 50 years and the creation of Argentine "Quasi-Par" bonds with 41 years until maturity, the achieved extension of the average life in restructurings is also expected to increase.

Table 3.2. Extension of duration and average life

	Pre-restructuring		Post-restructuring	
	Duration	Average life	Duration	Average life
Argentina	6.6	11.8	14.8	28.2
DomRep	3.6	4.8	5.6	7.6
Ecuador	2.7	3.3	9.4	20.3
Grenada	5.4	10.0	13.0	19.9
Moldova	0.0	0.0	3.6	4.4
Pakistan	0.3	0.3	3.7	4.5
Russia	6.9	19.5	9.6	27.6
Ukraine	0.6	0.6	3.6	4.5
Uruguay	5.0	8.5	7.4	16.9
Average	3.5	6.5	7.9	14.9

Source: Author's calculations. The duration given in the table is the weighted average of payment times, applying a 10% discount rate to ensure comparability. The average life is the duration of the principal payment. The measures given in the table are an average per country weighted by their respective issue size.

All restructurings featured special incentives for investors to attract participation; these are often referred to as "sweeteners", while here called "carrot" in contrast to "stick" features. Table 3.3 gives an overview, even if some of these features may appear to be common

---

short bond, a principal haircut of 30% was demanded; for the 20 year intermediate bond, a 20% haircut was deducted. Only the 30 year long bond was exchanged at par.

[41] See Section 4.2.7 for a definition of these measures.

or self-evident. Traditionally, past-due-interest has always been paid in cash, although this might change in the future. For example, Dominica offered cash payments only if bonds were tendered during early stages of the offering period. Repayments of a fraction of the notional in cash are only considered for issues which would originally have had matured already (which was the case in Moldova) or are due in the near future. Regarding the terms of the new claims, it is very popular to use bonds which start amortizing before their final maturity. This aims at an even cash flow pattern where principal repayments are less likely to cause liquidity problems.[42] Furthermore, the new terms can provide incentives such as increased coupons or some kind of bonus, like the GDP-linked warrant in the recent case of Argentina. Legal and regulatory incentives can take many forms. Market standard are cross-default and cross-acceleration clauses which exclude the original issues. The most favored investor clause, stating that the same conditions have to be granted to all investors, was designed to ensure that all participants in the exchange will benefit from any settlement under improved conditions. After being used by Argentina in 2005 for the first time, a cast-iron proof of its practical and legal effectiveness is, however, still missing. If the restructured bonds are issued by a body of higher seniority, this is called debtor upgrade. In Russia, the old Prins and IANs were issued by the state-owned Vneshekonombank while the new eurobonds are directly guaranteed by the Russian government. The use of trust-indentures is becoming more popular since it prevents future payments from being attached in the manner of the *Elliott vs. Peru* case.[43] Further legal provisions can offer insurance against repeated default, for example by reversing the haircut (which was achieved by issuing so called "contingent recovery rights" in the Ecuador restructuring) or making the new bonds putable (as included in the Russian restructuring). In most cases, regulatory sweeteners are only relevant to local investors. Argentina, for example, started accepting government bonds as liquidity requirement for local banks to prop up demand for the bonds, and allowed for accounting the bonds at par value. Gaining popularity, debt management clauses prescribe the issuer to buy back bonds on the market. These buy backs can be mandatory or contingent on the country's payment capacity (as in the case of Argentina).

Plainly spoken, "sticks" include all measures in an exchange to make the existing bonds less attractive. These can be unilaterally introduced

---

[42] This lesson has been learned especially by Ukraine: their new bonds started amortizing with the first coupon payment.

[43] See, for instance, Gulati and Klee (2001).

by the authorities (for instance by increasing the risk-weighting with
regard to the local bank's capital-adequacy ratios) or agreed upon by
the investors in exit consents. The most common exit consents remove
the cross-default and cross-acceleration clauses from the old bonds and
lift the listing requirement.

Table 3.3. "Carrot and stick" incentives of restructuring offers

Carrot features	Examples
Cash payment	
Accrued or past-due interest	almost all
Notional	Moldova, Uruguay
Terms of payments and bonuses	
Amortizing notional	Argentina, Pakistan, Ukraine
Coupon frequency increase	Ukraine
GDP-linked warrant	Argentina
Liquidity increase	Argentina, Pakistan, Ukraine
Legal incentives	
Cross-default clause excludes old bonds	DomRep, Uruguay
Aggregation clause	Argentina, Uruguay
Most favored investor clause	Argentina
Debtor upgrade	Pakistan, Russia
Trust indenture	DomRep, Ecuador, Uruguay
Default provisions	Ecuador, Russia
Regulatory incentives	
Liquidity requirement for local institutions	Argentina (April 2001)
Accounting at par value	Argentina (Nov. 2001)
Debt management provisions	Argentina, Dominica, Ecuador

Stick features	Examples
Exit consents	
Amendment of cross-default clause	Ecuador
Amendment of cross-acceleration clause	Uruguay
Amendment of negative pledge clause	Ecuador
Attachment prohibitions	Uruguay
Delisting	DomRep, Ecuador, Uruguay
Regulatory hurdles	
Increased risk-weighting	Uruguay

Source: Sturzenegger (2002), IMF country reports, prospectuses of offerings.
Features of bond exchange offers, making the exchange more attractive (so called
"sweeteners", here referred to as "carrots") and threatening holdouts (here referred
to as "sticks").

Majority voting has played a role in almost all restructurings to change the contractual terms of the old bonds. This was hardly recognized in the recent debate on collective action clauses. With regard to the restructurings analyzed in this study, all bond issues with majority voting were subject to either London or Luxembourg law. Thereby, majority voting was applied in very different ways. Moldova applied CACs to amend the terms of payment according to the restructuring offer. Since the TCW asset management company held 78% of the outstanding bonds, an agreement with this majority bondholder ensured the required 75%-majority vote. Uruguay used the CACs contained in its Samurai issue which presented the first use of CACs in Japan. The conditions of the original claim were amended by a two-third majority vote upon a minimum quorum of 50%. Because all other bonds issued by Uruguay did not include CACs and had to be exchanged, such an approach has been called "hybrid". However, the first country to actually apply such a hybrid approach was Ukraine. While its D-mark bond did not include CACs, Ukraine applied a quite innovative two-step mechanism to those bonds with CACs. In a first step, Ukraine invited the investors—mainly investment banks and hedge funds—to tender their bonds by granting an irrevocable proxy vote for the restructuring offer. Once the government collected sufficient votes, it called a bondholder meeting where the proxy votes were automatically cast in favor of modifying the terms of the old bonds. In a second step, all old bonds were exchanged for new ones. Other countries, like Pakistan and Russia, refused to invoke the contractual possibility for majority voting. Pakistan simply feared the calling of a bondholder meeting which might be conducive to investor coordination. It could be that the opponents of the restructuring were over-represented in such a meeting. To understand Pakistan's concerns it has to be kept in mind that Pakistan was the first country in which private sector involvement was strictly invoked by the IMF, creating much resistance among the investors. According to Sturzenegger (2002), Russia did not apply majority voting to avoid giving the restructuring an involuntary character. More appealing is the reasoning that the majority threshold of the Russian CACs was exceptionally high—95% for Prins and 98% for IANs.

Recent developments show that a participation rate close to 100% is not necessarily the conditio sine qua non for a successful restructuring. For this reason, defining an ex ante participation threshold is becoming less popular. In some cases, the participation threshold is dictated by the majority requirement from CACs anyway, as it was the case in Moldova. But in cases like Dominica, where the bonds were subject to a legal dispute, the circumstances are too uncertain to meet an ambitious threshold.[44] The pioneer restructurings—those of Pakistan, Ecuador, and Ukraine—have set a very high benchmark for realized participation, 98% on average. These benchmarks were not touched ever after. The 2005 mega-restructuring of Argentina has set a new and unglamorous mark with 76.1%. As a result, the country received massive critique from investors and banks and continues to struggle with holdout litigation. The IMF has pressed Argentina to offer reasonable compensation to holdout investors.

Some authorities considered a reopening of the exchange even after the official deadline or after the exchange was settled. The range reaches from an extension of the deadline (as in Pakistan), or a reopening of the exchange offer after the successful settlement of the first tranche (as in Ukraine), to the indefinite extension of the offer beyond the deadline (as in Dominica). This is some kind of courtesy to those investors who initially intended to hold out but were surprised by the high realized participation rate, and better decided to tender.[45] The reopening has to be considered as a mutually beneficial act, especially for soft restructurings with attractive terms. In the case of hard restructurings, a custom of reopening the offer appears like a real option to investors and would create incentives to hold out. This discussion, however, might be of limited endurance only. As majority voting turns out to gain dominance in facilitating orderly workouts, the future participation threshold for restructurings will be determined by the required majority of votes.

The limited involvement of investors—in other words those who actually have to accept the losses—in drafting the exchange offer has often enraged bondholders, especially retail investors. The recent closing of the Argentine restructuring offer fuels further disappointment. While the governments try to depict their negotiation strategy as cooperative, investors perceive an exchange often as an unilateral offer and feel like they are left out of the game. The truth is probably found somewhere in between. Formal negotiations, for instance, with some representative body of all bondholders, are undoubtedly the exception. There are two main reasons for this. First, it is often impossible to co-

---

[44] Actual participation in Dominica was hindered by the fact that the bonds were horizontally stripped into single zero coupon-like instruments and sold to a broad investor base. As a result, it is almost impossible to persuade all holders of the single zero coupon strips to tender their claims because investors' interests are very different and equality of treatment is hard to ensure.

[45] In practice it also occurred that some retail investors simply missed the initial deadline.

ordinate a great number of investors—some hundred thousands in the case of the restructured Argentine bonds. Against this background it makes sense to build bondholder groups, which actually were on the scene during the Argentine episode.[46] Yet they were not able to solve the coordination problem since they could not unite a critical mass of bondholders. The power of representation groups appears basically to be of a political kind, and therefore limited. Second, engaging in formal negotiations only makes sense to the issuer if the results are binding in any sense. This is the case when there are collective action clauses. If either the portion of bondholders represented at a meeting does not represent a sufficient quorum, or the competence of respective representative bodies do not allow for majority actions, it is hard to argue that bondholder meetings would make any sense except for showing some good faith.[47] In a hard restructuring, the rational government usually avoids formal negotiations due to the fear of creating a political stage for its opponents.

Informal or indirect negotiations have been involved in all restructurings. Indirect negotiations are facilitated by multilateral institutions, mainly the IMF along its basic principles about good faith negotiations.[48] Most restructurings—except for Argentina and Moldova—were accompanied by intact IMF programs under which an open dialogue is favored. Other than that, countries make use of informal contacts with bondholders and market participants. These range from open bondholder meetings and meetings with some large single bondholder to informal contacts with asset managers and traders through their financial advisers (so called "market sounding"). This feedback usually entails enough information to reach a conclusion about the bondholder's expectation regarding the restructuring.

This section has shown that certain patterns emanate from a wide choice of options in sovereign restructurings. The practice of sovereign workouts is under constant change, so some of the features will be subject to change in the future. Among those are the handling of past-due interest and legal innovations, mostly collective action clauses. Even without any centralized sovereign bankruptcy regime, the interaction

---

[46] To name a few, the Global Committee of Argentine Bondholders (GCAB), which acted as umbrella group in the end; the Argentine Bondholder Restructuring Agency (ABRA), mainly active in German speaking countries; and the Task Force Argentina (TFA) in Italy.

[47] Usually, the necessary quorum for a majority vote requires two-thirds of the face value to be represented at a bondholder meeting. See Dixon and Wall (2000), p. 144.

[48] See International Monetary Fund (2002).

between debtor, creditors, and the official sector definitely needs improvement. It has to be kept in mind that, as in any financial transaction of considerable size, any dissemination of information can have a high impact on markets and needs to be handled with care. The situation is especially complicated for sovereign authorities, given their high level of public exposure and the nature of diplomatic manners. A higher level of transparency with regard to consistent, reliable, and timely information—as postulated and promoted by the IMF—is without doubt desirable.

3.3.2 Resulting Present Value

What matters to investors most is the tradable value they receive in the very end. The main problem in calculating the expected net present value of the restructured claims is defining the appropriate discount rate. A number of different approaches are available, depending on the circumstances:

Benchmark curve by current bonds: In some cases prices of other performing bonds of the same sovereign issuer are available, since a restructuring might only include a certain class of instruments.

Benchmark curve by comparable sovereign: If there is some indication of the credit standing after completion of the exchange, a benchmark curve can be taken from countries with comparable circumstances.

Constant exit yield: A constant discount factor is applied.

For evaluating the Argentine offer in early 2005, analysts used all three approaches:[49] First, the Argentine US dollar denominated Boden bonds provided an Argentina-specific benchmark. This is disputable for many reasons. It neither includes the improved credit standing of the country after successfully closing the exchange nor reflects the fact that the Bodens belong to a totally different class of claims ruled under Argentine law and mainly owed to local investors. Second, the offer was also evaluated using benchmark curves from Brazil and comparable spreads from Ecuador. While indicative sources already suggested that Argentina would receive a B- rating after the restructuring, Brazil is rated BB- so that spreads are only a crude proxy. B- rated Ecuador, recently shaken by a crisis, has only two bonds of comparable characteristics outstanding which can only give a rough sketch of the term structure curve. Third, and by far the most frequently applied approach, a

---

[49] Analyses from Commerzbank, CSFB, Deutsche Bank, DrKW, JP Morgan, and some smaller banks were surveyed.

constant discount factor, most prominently 10%, is used. Less based on scientific reasoning, this flat discount rate can be an acceptable rule of thumb when reflecting overall market circumstances.

For the sake of a country-by-country comparison in this study, a constant exit yield is applied. To show the sensitivity of the present value to this choice, Figs. 3.2 to 3.7 illustrate the net present value (NPV) under different exit yield assumptions.[50] The value considers the total of cash payments and new bond value offered for one dollar of face value from the old securities plus accrued interest and past-due-payments, if any. The illustration shows that comparing the NPV result of restructurings is sensitive to the exit yield assumed. The sensitivity is higher the longer the maturity of the restructured instrument is.

Fig. 3.2. Net present value Argentina

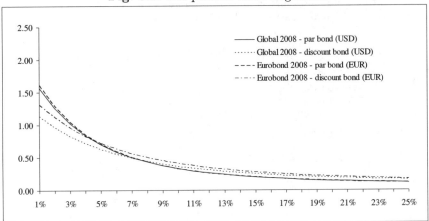

Source: Author's illustration. Net present value of of restructuring offer per one dollar of face in original securities under different exit yields.

A reality check is provided in Table 3.4, as the realized yields can be compared to the 10% exit yield assumption. In the first column, the "adjusted exit par yield" is defined as the theoretical yield at which the restructuring bundle—including new bonds and cash received per one unit of original face value—would equal par. At this yield, one dollar of face value in the original claim would be valued at one dollar after the restructuring. At current interest rates, this shows that the Argentine risk spreads would need to become negative for the restructuring to

---

[50] Only a selection of all different combinations between eligible and restructured instruments is shown.

Fig. 3.3. Net present value Ecuador

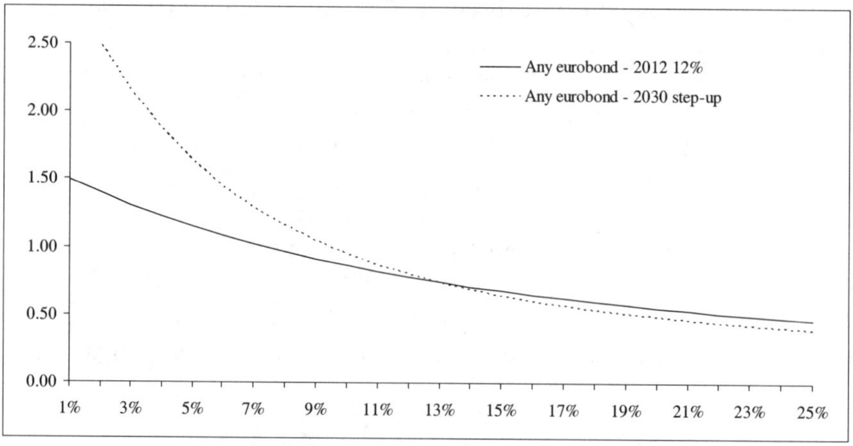

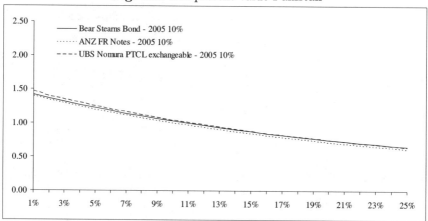

Source: Author's illustration. Net present value of of restructuring offer per one dollar of face in original securities under different exit yields.

Fig. 3.4. Net present value Pakistan

Source: Author's illustration. Net present value of of restructuring offer per one dollar of face in original securities under different exit yields.

equal par. The same applies to Russia. The Ecuador offer—the one with the third largest haircut—would be valued at par at a discount rate between 7.2% and 9.4%. Because spreads in Uruguay have not been extremely high before the restructuring, the exit par yields were comparably low around 6.7% to 8.5%. The difference of one percentage point in coupons results, for Ukraine, in significantly different exit par yields, around 16% for the dollar bond and 10% for the euro bond.

Fig. 3.5. Net present value Russia

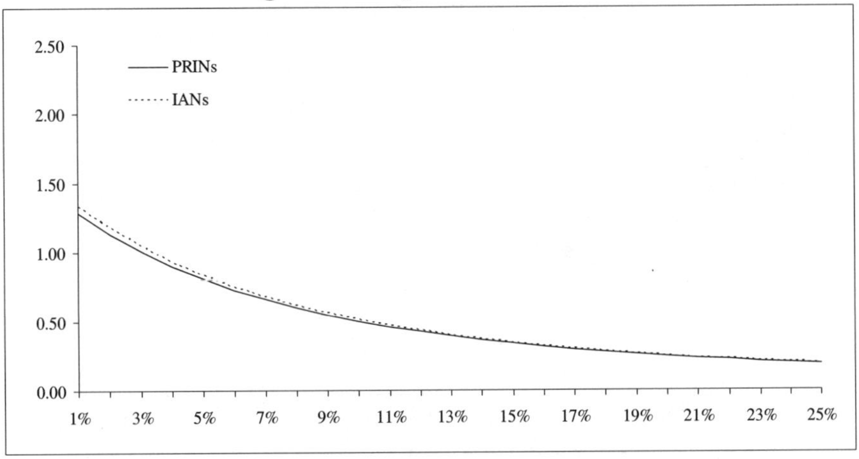

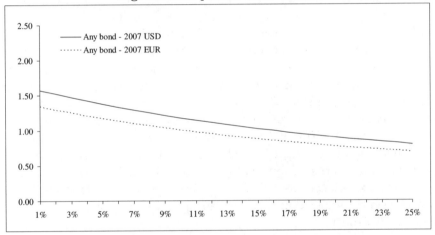

Source: Author's illustration. Net present value of of restructuring offer per one dollar of face in original securities under different exit yields.

Fig. 3.6. Net present value Ukraine

Source: Author's illustration. Net present value of of restructuring offer per one dollar of face in original securities under different exit yields.

Fig. 3.7. Net present value Uruguay

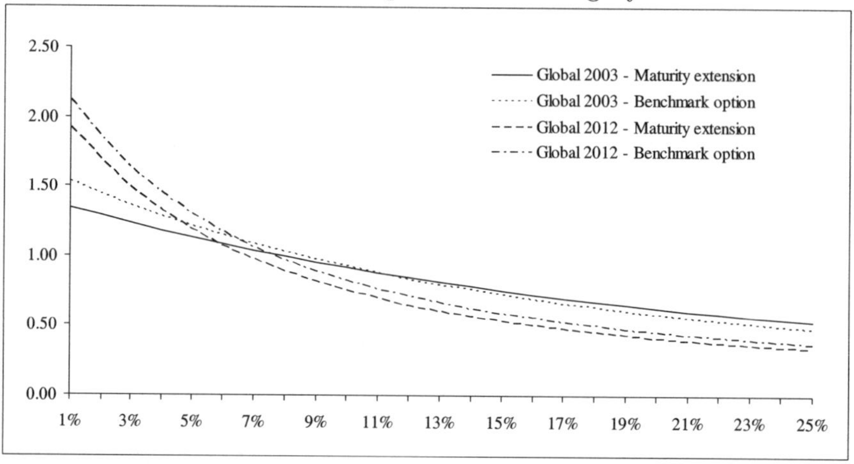

Source: Author's illustration. Net present value of of restructuring offer per one dollar of face in original securities under different exit yields.

In the following columns, the table shows the actual market yields from the first trading day (commonly called "exit yield", not to be confused with "exit par yield"), as well as market yields from six, 12, and 24 months after the settlement date. Apparently all realized market yields were high, indicating that the restructuring did not wipe out all distress phenomena immediately. Since the market yields are much higher than the corresponding par yields, all restructured bonds traded at significant discounts from par. Only in the softest restructuring seen so far, Uruguay, a 10% exit yield assumption proved ex post realistic. The same applies to Argentina only six months after the restructuring, paying tribute to the exceptionally low emerging market risk premia in 2005. The experience from these restructurings teaches that countries which recently underwent a restructuring have to expect yields above their market peers.

Table 3.4. Comparison of exit yields

Instrument tendered	Instrument received	Adj. exit par yield	First trading day	6M	12M	24M
Argentina						
Global 2008	Par bond (USD)	3.1%	8.4%	8.3%		
	Discount bond (USD)	1.8%	9.3%	8.5%		
Eurobond 2008	Par bond (EUR)	3.2%	8.1%	8.1%		
	Discount bond (EUR)	2.7%	9.0%	7.8%		
Ecuador						
Any Eurobond	2012 12% USD	7.2%	21.2%	17.7%	19.9%	22.2%
	2030 step-up USD	9.4%	20.1%	17.9%	19.6%	20.3%
Pakistan						
Bear Stearns Bond	2005 10% USD	11.0%	20.7%	26.4%	27.7%	15.2%
ANZ FR Notes	2005 10% USD	10.0%	20.7%	26.4%	27.7%	15.2%
UBS Nomura PTCL exchangeable	2005 10% USD	11.3%	20.7%	26.4%	27.7%	15.2%
Russia						
Prins	2010 8.25%/2030 step up	3.1%	15.1%	16.8%	14.4%	10.1%
IANs	2010 8.25%/2030 step up	3.4%	15.1%	16.8%	14.4%	10.1%
Ukraine						
Any bond	2007 11% USD	15.7%	22.4%	17.9%	19.9%	10.3%
	2007 10% EUR	10.0%	21.5%	16.0%	16.9%	9.8%
Uruguay						
Global 2003	Maturity extension	7.8%	18.1%	17.4%	15.3%	7.4%
	Benchmark option	8.5%	11.9%	9.8%	10.9%	7.6%
Global 2012	Maturity extension	6.7%	12.8%	10.5%	10.3%	
	Benchmark option	7.7%	11.8%	9.7%	10.3%	

Source: Author's calculations.

The table displays exit yields for selected bonds which underwent a restructuring. The adjusted exit par yield is defined as yield at which all components received in the restructuring equal the par value of the original instrument. The following columns show actual market yields on the first trading day, as well as after six, 12, and 24 months, where available.

For comparing the NPV reductions resulting from restructurings, different approaches are possible. The popular term "haircut" mostly regards the exchange ratio, i.e. the change in face value. This cannot be the whole story, since maturity and coupons are other important determinants of the bond value.[51] Therefore, it is more advisable to compare the present value of a bond's promised cash flows, be it by looking at market prices or by discounting future payments.

A new way to illustrate the character of a restructuring is to plot the pre-restructuring value against the net present value received after a restructuring. To ensure comparability, Figure 3.8 uses a constant exit yield of 10% for all countries although the overall picture is insensitive with regard to the exit yield applied. This type of illustration makes it possible not only to assess the net present value implications of the restructuring, but also to account for the differences in trading value prior to the credit event. Since the characteristics of single bonds—such as maturity and coupon—can differ widely, inter-creditor equality should also mean that the restructuring accounts for the difference in trading value. As the figure unveils, two extreme cases are observable. In Argentina, as one extreme, all investors were basically offered the same terms of the restructuring regardless of the initial bond characteristics. The 2003 11% Italian lira bond, trading at par in June 2001, was treated the same way as the 2031 12% capitalizing US dollar bond, trading at 76%. In Figure 3.8, this results in a scatter band around a horizontal line. Argentina, Ecuador, Pakistan, and Ukraine offered similar terms to different bondholders, resulting in points on an almost horizontal line.[52] The other extreme is Uruguay. For most eligible bonds, the restructuring offered a direct counterpart with comparable characteristics except for a longer maturity. In the figure, such an approach results in points lying on an upward-sloping regression line.[53]

When interpreting Figure 3.8, two caveats have to be kept in mind. First, a flat discount factor does not reflect an usually downward sloping term structure. This can impact, for instance, Uruguay since the short-term bonds traded at a significantly higher yield right after the restructuring (see Table 3.4). Second, the choice of the period of six

---

[51] Sturzenegger and Zettelmeyer (2005) show that NPV losses are mostly unrelated to principal haircuts.

[52] Not considered here are Brady bonds which received differential treatment in Ecuador and Argentina.

[53] A regression of the points representing a selection of Uruguay exchange options, the intercept is 0.65 and the slope coefficient is 0.41. If the intercept is set to zero, the slope coefficient is 1.75.

Fig. 3.8. Market price before and NPV value after the restructuring

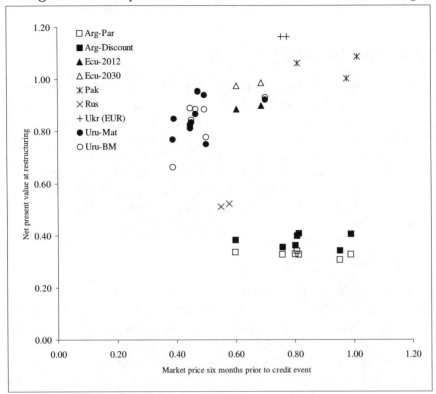

Source: Author's illustration.
Comparison of market prices six months before the credit event—mostly default
or restructuring, whatever is first—and the net present value received in the
restructuring at a discount factor of flat 10%.

months before the credit event is somehow arbitrary. It is debatable
at which point in time the market participants could foresee the restructuring. Although this depends on the individual circumstances,
a six month time span before the respective credit event is a reasonable choice. At this point in time, crisis symptoms are already priced
into bonds, but there is not yet certainty about the restructuring and
its terms. Actually, bond prices in Uruguay were lower six months before the exchange, but recovered in the following months since fears
of a hard restructurings decayed and the cooperative and soft nature
of the restructuring became clear. Similar effects can be discussed for
each country although the base message of the plot remains, namely

the fundamental difference between an equalizing restructuring (like Argentina) and a differentiating restructuring (like Uruguay.)

3.4 Lessons for Investors

Restructurings often mean different things for investors than they do for the countries themselves. While the economic relief granted by a restructuring is one story, the effects on the international capital markets can be quite a different one. The question is how well investors with commercial interests (in contrast to concessional lending by the official sector) actually fare in restructurings. This section sheds light on the returns on an investment realized during recent restructurings, and draws conclusions with regard to what recovery value should be expected.

3.4.1 Investor Returns

While a credit event like a restructuring with or without previous default might be foreseen by market participants, it is worth looking at the resulting returns for investors during a longer period of time. Figure 3.5 shows returns during different holding periods.

The credit event is defined as the date of the actual default or the opening date of the restructuring offer, whichever comes first.[54] This date marks a clearly recognizable milestone in the restructuring process and is less ambiguous than the beginning of the renegotiation process.

Using these milestone dates, table 3.5 shows the annualized return for different kinds of investment strategies during restructurings: first, a buy-and-hold investor who bought six months before the credit event and holds on until 24 months after closing of the restructuring;[55] second, an investor who bought six months before the credit event but sold the issues at the opening day of the exchange; third a vulture investor who bought original claims at that day, tenders it into the restructuring, and holds the new instruments until 24 months after the exchange became effective; fourth, the simple return of an investor who bought the old instrument at the date of the exchange opening and sold the new instrument on its first trading day. The last, fifth column shows an investor who bought the new issue on its first trading day and holds it

---

[54] As default date this study chooses either the day of the official default declaration (as in Argentina), or the end of the idle grace period for pari passu ranking liabilities.

[55] The respective time horizon for Argentina is six months.

for 24 months. Keep in mind that financial distress in all of the returns was evident even six months before the actual credit event.

This analysis differs from the view taken in Sturzenegger and Zettelmeyer (2005). The authors claim that an ex post analysis which compares the market value of the original bond with the resulting market value of all cash and claims received after the restructuring is, at its most basic level, a test of market efficiency; The rational market value right before the restructuring should be nothing else than the expectation about the terms of the restructuring and the participation rate. Therefore, such a comparison would mainly measure how well the market was informed about the upcoming restructuring and whether the market reacted rationally to this information. It has to be borne in mind though that investors usually prefer to tender at the latest possible moment which might be weeks or even months after the opening date. It is usually during this time span that new valuable information about the fundamental situation, as well as about the level of participation and the holdout litigation strategy, arrives.

However, alternative approaches can pose other disadvantages. Sturzenegger and Zettelmeyer (2005) use exit yields of the new instruments to price the original claim and compare the post-restructuring market value with the pre-restructuring NPVs at those yields. A number of practical complications arising from this approach are already addressed in their paper.[56] The main problem of such an approach, however, lies in the fact that a restructuring is intended to improve the fundamental creditworthiness of a debtor. For this reason, using post-restructuring yields distorts the picture exactly in those cases where the restructuring has successfully contributed to lower spreads. If a very hard restructuring (like Argentina) helps to reinstall the long-term solvency and significantly reduces the risk spreads, this approach exaggerates the perceived "haircut" since the pre-restructuring NPV would become unrealistically high in comparison to the pre-restructuring bond price. Considering these pros and cons, this study chooses to compare gains and losses from restructurings by looking at realized (in contrast to theoretical) investment returns during different holding periods.

---

[56] The new bonds might have a different maturity structure or there are not enough bonds to derive the structure of the term structure from them. Furthermore, it can be problematic to compare pre- and post-restructuring yields when the restructuring includes a change in currency, seniority, jurisdiction, or some financial innovation such as GDP-linked bonds.

Table 3.5: Annualized returns during restructurings

Instrument tendered	Instrument received	Buy and hold	In exchange Sell old	In exchange Buy old	Rent of exchange	Buy new
Argentina						
Holding period (years)		*4.4*	*3.6*	*0.8*		*0.4*
2008 15.5% USD	Par bond (USD)	-15%	-23%	26%	**16%**	9%
	Discount bond (USD)	-16%	-23%	20%	**6%**	22%
2006 11.25% EUR	Par bond (EUR)	-18%	-28%	49%	**28%**	17%
	Discount bond (EUR)	-17%	-28%	63%	**32%**	30%
2008 Step-up EUR	Par bond (EUR)	-14%	-25%	56%	**33%**	17%
	Discount bond (EUR)	-13%	-25%	71%	**37%**	30%
2017 11.375% USD	Par bond (USD)	-15%	-22%	29%	**18%**	9%
	Discount bond (USD)	-14%	-22%	34%	**16%**	22%
2018 capitalizing USD	Par bond (USD)	-14%	-21%	25%	**15%**	9%
	Discount bond (USD)	-15%	-21%	17%	**5%**	22%
2029 8.875% USD	Par bond (USD)	-9%	-22%	78%	**52%**	9%
	Discount bond (USD)	-9%	-22%	78%	**45%**	22%
2003 7% CHF	Par bond (EUR)	-19%	-33%	93%	**57%**	17%
	Discount bond (EUR)	-19%	-33%	83%	**45%**	29%
Ecuador						
Holding period (years)		*3.5*	*1.2*	*2.3*		*2.0*
2002 11.25% US$	2030 step up US$	-8%	-44%	18%	**45%**	1%
	2012 12% US$	-6%	-44%	22%	**36%**	8%
2004 FRN US$	2030 step up US$	-5%	-18%	3%	**5%**	1%
	2012 12% US$	-3%	-18%	6%	**-1%**	8%
Pakistan						
Holding period (years)		*2.6*	*0.5*	*2.1*		*2.0*
1999 11.50% US$	2005 10% US$	-3%	-38%	8%	**-6%**	12%
2000 FRN US$	2005 10% US$	-3%	n.a.	n.a.	**n.a.**	12%
2002 6% US$	2005 10% US$	7%	-45%	26%	**28%**	12%

(continued on next page)

Table 3.5: Annualized returns during restructurings (continued)

Instrument tendered	Instrument received	Buy and hold	In exchange Sell old	In exchange Buy old	Rent of exchange	Buy new
Russia						
Holding period (years)		*4.2*	*2.1*	*2.1*	*2.1*	*2.0*
IAN	2010 8.25%/2030 step up	-1%	-24%	30%	**10%**	26%
Prin (US$)	2010 8.25%/2030 step up	0%	n.a.	n.a.	**n.a.**	25%
MinFin III	MinFin VIII	34%	n.a.	n.a.	**n.a.**	29%
Ukraine						
Holding period (years)		*2.8*	*0.6*	*2.2*	*2.0*	*2.0*
2001 16% DEM	2007 10% EUR	11%	-13%	19%	**-14%**	30%
2000 14.75% US$	2007 11% US$	12%	-13%	20%	**-13%**	30%
Uruguay						
Holding period (years)		*2.6*	*0.5*	*2.1*	*2.1*	*2.0*
2003 7.875% USD	Maturity extension option	15%	84%	3%	**-29%**	23%
	Benchmark option	17%	84%	5%	**-11%**	12%
2006 8.375% USD	Maturity extension option	31%	26%	32%	**40%**	14%
	Benchmark option	31%	26%	33%	**44%**	13%
2008 7% USD	Maturity extension option	31%	35%	30%	**33%**	15%
	Benchmark option	34%	35%	33%	**45%**	13%
2009 7.25% USD	Maturity extension option	33%	24%	35%	**35%**	19%
	Benchmark option	33%	24%	35%	**47%**	13%
2010 8.75% USD	Maturity extension option	34%	25%	37%	**47%**	15%
	Benchmark option	37%	25%	40%	**60%**	13%
2012 7.625% USD	Maturity extension option	30%	39%	28%	**25%**	17%
	Benchmark option	36%	39%	35%	**48%**	13%
2027 7.875% USD	Mat. extension/benchmark	39%	47%	37%	**48%**	15%
2005 7% EUR	Maturity extension option	25%	21%	26%	**32%**	12%
	Benchmark option	24%	21%	25%	**28%**	12%
2011 7% EUR	Maturity extension option	39%	59%	35%	**52%**	11%
	Benchmark option	29%	59%	23%	**23%**	12%
2009 7.875% USD	Maturity extension option	33%	25%	35%	**38%**	17%
	Benchmark option	34%	25%	36%	**52%**	13%

Source: Bloomberg, Datastream, author's calculations. n.a.: data not available.

Annualized returns for Argentina in Table 3.5 reflect the extraordinarily long period between default and restructuring. This makes losses on the old instruments appear less severe than the actual price drop suffered during the second half of 2001. After some legal skirmishes, the new bonds commenced trading in June 2005, immediately realizing considerable gains to those investors who tendered their bonds.

Buy-and-hold returns for Ecuador have been negative despite declining overall spreads as represented by the EMBIG. Investors realized the largest losses when selling the old instruments right before the exchange. The exchange itself was an unexpected success, rewarding those who bought old instruments and tendered them with a simple return of up to 45% for the fixed coupon bond, mainly due to the high cash component from past due interest. During the next two years, the return on the new instruments was slack, since Ecuador did not achieve a substantially improved credit standing and was downgraded by S&P from B- (which was received after the restructuring) to CCC+ in April 2001.

In Pakistan, bond prices started to tumble 18 months before the actual restructuring took place. Before that, bonds traded around par—a level which was not reached again until two years after the exchange. The restructuring period shows a very mixed picture. As prices did not converge (although the eurobond was putable in February 2000) but the exchange ratio was almost similar for all instruments, the exchange itself had a very uneven impact. The price recovery of the new instrument amounted to about 12% annually. S&P maintained its B- rating during this period.

The situation in Russia is marked by the big bang of the Russian crisis and a remarkable recovery which continues until today. The calculations use the default of the Prins on December 2, 1998 as date of the credit event because Russia defaulted on several instruments in succession. After the restructuring, all instruments gained from Russia's fast recovery, increasing by about one quarter in value annually. Overall buy-and-hold returns are considerable for the MinFin III bond since no haircut was required.[57]

Market prices for Ukrainian bonds are only available for the 2000 dollar eurobond and the 2001 16% D-mark bond. Bond prices suffered only moderately after the restructuring announcement since Ukraine's creditworthiness was under dispute for some time. Due to the cut in coupons, the rent of the exchange turned negative although the fol-

---

[57] While IANs and Prins were issued under English law, the MinFins obey Russian legislation and do not include cross-acceleration clauses.

lowing price recovery was remarkable. While investors who bought the new issue gained 30% annually, buy-and-hold investors could realize an annual 11% return throughout the crisis.

As can be seen from the returns in the six month period before the exchange opening, the preventive Uruguay restructuring was perceived as a relief after a period of uncertainty: the open communication policy helped prices to recover even before the closing date. Investors who were holding old issues throughout the six months prior to the offer realized an annualized return of more than 30% on average. The restructuring itself offered a return of another 30% on average (except for the 2003 global bond which had a remaining life of six months). After the transaction, recovery of bond prices was moderate and remained close to the overall market returns of emerging market bonds.

Overall, the table proves evidence that buy-and-hold investors who bought six months before the credit event (especially when the country is already in distress) did not have to accept losses in market value terms. Investors may also have benefited from the fact that so far all restructurings became successful transactions with high participation rates and robust market confidence. The simple return of a restructuring exchange—buying and tendering the original instrument and selling the new one on its first trading day—has been very significant in all cases. When looking at the return of an investor who sold on the credit event (which can be compulsory for some institutional investors) it seems that they have already weathered the worst. Once in distress, there is no evidence that the market condemns a restructuring.

3.4.2 Modeling the Recovery Value

By describing the customs of ad hoc sovereign restructurings, the above has identified some patterns which should be properly reflected in a model to value sovereign bonds. The objective of this exercise is to correctly incorporate the usual conditions of sovereign restructurings in the pre-default value of sovereign bonds. Models of credit risk achieve this by adjusting the risk-free bond price by the probability of default, assuming some recovery value to be paid upon such a credit event. Bringing models of recovery value in line with the stylized facts is the focal point of this section.

Up to now, this issue has not been cultivated much in sovereign risk valuation. As pointed out in Chap. 2, indicators of country risk are used in empirical studies to determine the risk spread, or the yield, of the respective country. This measure applies uniformly to a country according to an index of credit risk, such as the average country risk

spread measured by the EMBIG. This gives little guideline for pricing
pricing single bond issues.

When looking at single bonds or the term structure of risk premia,
market practitioners mostly assume an exogenously and invariant recovery value, either as a fraction of the pre-default market value or as a
fraction of face value. In contrast to academia, however, market practitioners are aware of the distinction between bonds trading on a "yield
basis" or a "price basis". The former refers to a set of bonds subject to
the same default risk and presenting a smooth term structure of yields.
This is the normal case in the absence of significant default risk and
gets reflected by modeling recovery as fraction of market value. When
bonds are traded on a price basis, however, the yield curve is little infor-
mative as the bonds trade at roughly the same price, mostly far below
par. The resulting yield term curve is inverted and often jagged. This
occurs upon fears of a pending default which is expected to lead to an
equalizing restructuring as discussed in the previous section. Models of
credit risk resemble this situation by assuming a recovery of face value
(RFV).

These alternative concepts of modeling recovery present a starting point for reevaluating the recovery expectations. Theory and practice have been focused on the widely applied recovery of market value
(RMV) concept.[58] In this framework, the expected recovery value is a
fraction of the survival-contingent market value under the same risk-
neutral probability measure.[59] The beauty of this assumption lies in
the fact that the recovery rate and the default intensity merge into one
number, being the product of both. The term risk spread often refers to
this product and is often compared to other spreads like credit default
swap (CDS) spreads, although these obey different recovery assumptions.[60]

Although simple and popular, this concept has both pros and cons
when applied to risky sovereign bonds. In one sense, the RMV concept
often results in a negative correlation of the default probability and the
recovery value. When spreads rise in financial distress, it is correct to
assume that this originates in a rise in the default probability. Since this
lowers the survival-contingent market value, the RMV scheme assumes
ceteris paribus that the recovery fraction per one unit of face value is

---

[58] See Duffie and Singleton (1999), p. 691.

[59] See Sect. 4.4 in Chap. 4 for a more formal presentation.

[60] The payoff of a CDS contract given default is the face value minus the post-
default value of the cheapest-to-deliver bond, thereby mimicking the recovery of
face value scheme. See Chap. 6.

declining as well. Thinking in terms of such a recovery of face value concept, this results in a negative correlation between default probability and RFV recovery fraction which is supported empirically. In the case of sovereign defaults one could correctly presume that the later in a crisis a default takes place, the lower the recovery value would be. Conversely, the RMV concept has a major shortcoming for two reasons. First, the RMV concept does not account for equal treatment in the sense of pari passu ranking after default.[61] The underlying economic reasoning is as follows. In the case of default, the only contractual lever investors can pull is acceleration.[62] Due to the common cross-acceleration clauses in pari passu ranking bonds, all these bonds are immediately repayable at par if a certain minority of all bondholders votes for acceleration. Other bond characteristics like maturity and coupon (which are responsible for the bonds to trade at different prices) lose their relevance.[63] From a legal stand point, all bonds should be treated equally and receive compensation according to their respective par value. In a restructuring, the corresponding approach would exchange any number of different bonds (with different coupons and maturities) into one new issue. This has been the case, for instance in Pakistan. Argentina and Ukraine are additional examples. Second, the RMV approach does not provide a lower bound for bond prices in distress which poses problems when the credit event is imminent. Once default appears to be inevitable and the default probability is therefore extremely high, the model bond price would collapse to close to zero under the RMV framework. But empirical observations show that sovereign bond prices have some lower bound even when close to default. This arises from the investors' expectation of receiving at least some lump sum even in the worst case. Since it does not make much sense to assume that the recovery fraction of market value would then increase to some multiple of one, such a lower bound can better be modeled by assuming a certain recovery fraction of face value.

The above is not only true for traditional cases of credit events like a payment default followed by a hard restructuring. While voluntary debt swaps (as argued in Sect. 3.2.2) have generally no specific price impact and, ideally, do not constitute a credit event, soft restructurings do. Considering them as a valuation-relevant event (similar to hard restructurings) is appropriate for three reasons. First of all, a soft re-

---

[61] See Duffie and Singleton (1999), p. 702.

[62] Upon an event of default (i.e., after the grace period), each individual holder is entitled to declare the principal amount to be due and immediately payable.

[63] Except for due but not yet paid interest.

structuring is comprehensive. While any selection of bonds can become eligible in voluntary debt swaps, a restructuring presents one comprehensive transaction including all bonds of equal ranking. Secondly, a soft restructuring is exclusive. Any restructuring marks the end of a story and is the last measure taken in a crisis. Up to now, a soft restructuring has been an appropriate instrument to halt a debt crisis and therefore makes another debt swap or a hard restructuring in the near term unnecessary. If a restructuring is not ill-designed, a soft restructuring serves its purpose of granting short-term cash flow relief and ensuring medium term debt sustainability.[64] Lastly, a soft restructuring bears some costs to the bondholder by definition. As is evident from Table 3.2, a soft restructuring usually extends the maturity structure. In some cases, a reduction in coupon or some grace period is also included which will not be made up by sweeteners. In terms of bond valuation, this affects the inputs into the bond value formula. Even if markets might react positively to a soft restructuring and market prices recover, any kind of restructuring features a credit event and bears some NPV implication.

Figure 3.8 provided evidence from past restructurings which enforces the distinction of RMV and RFV approaches. An equalizing restructuring offer (like Argentina) represents an RFV-type restructuring as holders of all bonds are offered the same exchange options, irrespective the original bond they hold. This results in points scattered around a horizontal line in Fig. 3.8. Differentiating restructuring offers, in contrast, take the terms of the original bond, as reflected by its maturity and coupon etc., into consideration. As these differences in bond characteristics lead to different bond prices, a differentiating restructuring would produce points around a significantly upward sloping line in the graph. By means of a simple regression, Table 3.6 presents supporting evidence for this notion from Argentina and Uruguay. The regression uses the post-restructuring NPV value as dependent variable which is regressed on a constant and the pre-restructuring market value.[65] On the one hand, the result supports the view that Argentina, the toughest restructuring seen so far, obeys an RFV scheme of recovery. On the other hand, Uruguay, the softest restructuring so far, was not a purely RMV-like restructuring (as might be supposed when looking at

---

[64] Of course, there is no theoretical obstacle to repeated restructurings.

[65] This regression uses the data of all restructured bonds (excluding Bradies) for which clean market prices have been available and weights them equally. The low number of observations for Ecuador, Pakistan, Russia, and Ukraine make it impossible to derive meaningful numerical results.

the terms of the exchange) but suggests a mixture of both RMV and RFV concepts.

Table 3.6. Regression of recovery values

Country	Intersect	Slope	df	R^2
Argentina	0.38***	-0.06	12	15%
Uruguay	0.53***	0.58***	18	36%

Source: Author's calculations. Regression of the net present value at a 10% discount rate offered in restructurings on a constant (mimicking the RFV scheme) and the respective clean bond prices (mimicking the RMV scheme) six months prior to the credit event.

These observations and findings are neither conclusive nor irrefutable. Nevertheless they make a strong point which can be underpinned by rational arguments. Soft restructurings are already the predominant way (and will be applied even more frequently) to ease and resolve upcoming debt problems. Although bond prices suffer from the accompanying economic circumstances in these situations, they do not collapse, and price differences are maintained. A soft restructuring offer will be designed in a way that it leaves the investor options to choose from and will respect the differences in the original claims in order to ensure high participation throughout all single bond issues and avoid holdouts. Uruguay is the best example.

However, any restructuring will present a lower bound for the recovery of par value where the notional amount is the creditor's most indisputable claim. Coupons, embedded options and other contractual benefits can get nullified in case of acceleration. Hard restructurings are likely to align the compensation with the initial notional claim investors hold. Argentina is the best example for this case, but basically all restructurings bear resemblance to this fact by equally honoring compensation for the notional claim of each bond. As the above regression has shown, this can be said about Uruguay to a certain extent as well.

3.5 Concluding Remarks

The expansion of sovereign issuers from emerging countries into the international capital markets requires a different attitude towards resolving debt problems. Two decades ago, bank loans and bilateral debt dominated the picture and commercial creditors aimed at restructurings which preserved the book value of their claims. Today, investors of listed bonds strive for a quick turn around of the distress situation in order to facilitate trading liquidity and a rebound of secondary market prices.

This has led to the distinction of liquidity crises in which a collective action problem of short-term refinancing needs to be overcome, and solvency crises which require a contribution from all stakeholders. Any path to crisis resolution is undoubtedly paved by the strategic behavior of multiple parties in a political process. However, holdout and litigation can be expected to play a smaller role in the future. The increased inclusion of collective action clauses in bond contracts will deprive minority groups of the power they currently enjoy. Preventing vulture investors from blocking sovereign restructurings seems to be the will of policy makers. Argentina is currently testing this stance, and more countries can be expected to follow this route once solvency problems emerge. From an investor's point of view it will become more desirable than ever to address debt problems at an early stage.

Besides establishing a pattern of resolving debt problems created by international sovereign bonds, this chapter describes the technicalities of restructuring offers. A look at the realized recovery values yields a new view on the meaning of inter-creditor equality within one class of equal ranking claims. It has always been the case that different bonds—possibly denominated in different currency, underlying different legislation, or held by different investor groups—are not priced homogeneously. While inter-creditor equality forbids preferred treatment for particular investor groups, honoring initial price differences in a restructuring has not been seen as a violation to the equity principle. Past restructurings are found to show a range from equalizing to differentiating recovery values.

Soft restructurings, which should be expected to dominate future resolution efforts, try to avoid a coercive character and are therefore more likely to offer attractive terms. Issuers might also consider such offerings to consolidate their debt and exchange existing bond contracts for new ones which include collective action clauses. Notwithstanding such transactions, the necessity (and their political prerequisites) for hard restructurings might arise. In this case, investors should be prepared to accept decisive "haircuts" but can hope for a significantly improved credit standing afterwards. However, as bond prices usually lose most value before the initiation of a restructuring, the analysis reveals that a distressed exchange itself has resulted in a remarkably positive return on investment in most cases.

From the vivid discussion about sovereign debt restructurings it is likely that these customs may undergo considerable changes in the future. However, the current setting already bears a direct impact on the way default and recovery risk influence sovereign bond prices and differs from the way corporate credit risk is modeled. This chapter has mapped the stylized facts from recent restructurings to the two dominant ways of modeling recovery, the RFV and RMV concepts. The following Chap. 4 builds on this by providing a mathematical formalization and suggesting a recovery model incorporating a mixture of both the RFV and RMV concepts. The resulting bond value model is empirically applied on sovereign bonds in Chap. 5 and sovereign credit default swaps in Chap. 6.

4

Modeling Sovereign Default Risk

The crucial question every investor asks before buying or selling a sovereign bond is whether the current market price is fair or whether it is supposed to rise or fall in the future. Every investor on his own implicitly applies some kind of bond valuation model. Such a model can vary from pure speculation about the markets' response to upcoming events, to an all embracing model of the global economy.

While the theoretical literature is interested in the development of sound models which replicate logical relationships and derive implications from underlying assumptions, the empirical literature looks at the real world outcome—the bond price. This is the aggregation of individual investors' calculus and their investment decision, serving as input for this empirical study. This empirical study will try to disaggregate the observed, revealing how a representative investor could have evaluated a sovereign bond.

This disaggregation, in turn, relies heavily on theoretical models of the pricing of credit risky claims. Any given result is conditional on the model applied. There is no guarantee that this model is "true". Furthermore, empirical work can only produce evidence about factors of actual influence. A well designed empirical study faces the following challenges: First, the underlying model needs to be a reasonable replication, both theoretically and intuitively appealing, of investors' calculus. Second, model and data need to fit together, providing a reality check on whether the empirical model represents a reasonable approximation of the "true" model and reflects the most relevant factors of influence.

What does this mean with regard to this study? The real world outcome researched here is the bond prices from sovereign issuers in emerging markets. These prices are the result of trading activity which emanates from individual investors' decisions to buy or sell. The trading price represents the sum of the expected present value of future cash flows from the respective bond. This is influenced by the choice of the discount factor and the occurrence of a default or restructuring. Finally, the question arises whether the applied inputs to the bond value were chosen appropriately according to the sovereign's economic capacity and willingness to fulfill the contract.

This chapter addresses the issue of how to accommodate all these individual components in one model for the bond price. In accordance with so called "reduced form models", unobservable, purely constructive variables are used for calculating measures like the default probability and the recovery value from restructurings or default. These measures, best suited to value bond contracts, are then estimated in Chapter 5.

The objective of this chapter, however, is to develop a sound model along two conditions. First, the model needs to reflect the particularities of sovereign bonds such as high coupons and long maturities and several bond issues outstanding sharing the same seniority but showing totally different characteristics. Second, the model needs to remain tractable for the empirical estimation. While it is theoretically possible to consider all potential influences on the bond price in the model, the empirical researcher must always balance the trade-off between the empirical richness of the model and its computational manageability. Specifically, the available empirical data needs to provide sufficient information to calibrate the model, otherwise the estimation will not converge and results will not prove robust.

With this caveat in mind, the structure of this chapter is as follows. Section 4.1 offers a short review of the literature specific to credit risk modeling and bond valuation. Section 4.2 introduces the mathematical foundations of analyzing bonds. The following section, Sect. 4.4, looks specifically at modeling the recovery value within the previously described frameworks of credit risk. These sections have been designed to suit a less technical audience as well and lead beyond the scope essential for understanding the empirical sections following in Chaps. 5 and 6. Finally, Sect. 4.5 outlines the empirical strategy utilized to implement the model of sovereign bond valuation in order to derive implied default and recovery parameters in the next chapter.

4.1 Literature Review

The following gives a short review of the development of the credit risk literature. While this review is not comprehensive, it offers many suggestions for further reading and provides a glimpse of how research addresses the valuation of fixed income instruments and credit derivatives.[1]

Any valuation approach of future cash flows uses a discount factor to determine today's equivalent value of future payments, namely, the present value. The literature on term structure modeling establishes consistent patterns of discount rates for a continuum of maturities, which can be calibrated to observed interest rates. The seminal contributions by Vasicek (1977), Cox et al. (1985b), Brennan et al. (1979), and Brennan and Schwartz (1982) have introduced stochastic processes to describe the shape of the term structure. A large body of today's asset pricing literature is based on these findings. Ho and Lee (1986) and Hull and White (1990) have introduced arbitrage-free models which can be fit to today's term structure of interest rates without creating arbitrage opportunities under the assumption that investors are risk neutral. Models with more than one stochastic factor were put forward by Duffie and Kan (1996) and Hull and White (1994) which allow for a richer pattern of the volatility structure than the previous one-factor models. The volatility term structure of the interest rates is thereby hump-shaped with a declining interest rate volatility in the distant future.

The credit risk literature draws heavily on these approaches, as will become apparent in the following sections. In general, credit risk is considered to be the risk of contractual failure to pay. This definition is very narrow. As Chap. 3 has shown, sovereign bond contracts are subject to renegotiations without necessarily accumulating arrears beforehand. Yet this does not contradict the general concept of "default" in the literature on credit risk pricing which assumes default to institute a certain point in time at which the original contract is surrendered and replaced by some recovery value.

Two major approaches, structural and reduced form models, have emerged from the literature on corporate credit risk on how to model the probability of a credit event at a certain time.

4.1.1 Structural Models

Structural models were initially based on the notion that corporate debt can be interpreted as a short put on the firm value. Also referred to as the "firm value approach", default is triggered when the firm

---

[1] Common text books on this topic are Ammann (1999), Bielecki and Rutkowski (2002), or Schönbucher (2003).

value slips below its contractual debt obligations. Merton (1974) models default as a situation in which assets to pay down the maturing debt are insufficient, drawing on the common option pricing framework by Black and Scholes (1972). The first passage time approach abstracts the initial idea by defining the default as the first point in time when some measure of the repayment capacity (such as a company's cash flows or its assets-liability-ratio, not exclusively the firm value) drops below some threshold. A large number of other authors have devoted their work to this approach, like Black and Cox (1976), Brennan and Schwartz (1977), Geske (1977), Brennan and Schwartz (1980), Leland (1994), Longstaff and Schwartz (1995), Leland and Toft (1996), Zhou (1997), and others.

Applying structural models on sovereigns is not straight forward since a sovereign country is not assigned an entity value. By the use of other indicators of debt capacity, the structural approach was transformed to suit sovereign credit risk. For instance, Kulatilaka and Marcus (1987) model GDP as a stochastic process which determines debt capacity. Debt servicing is seen as a drag on future GDP growth so that a country decides to default when the costs of the default sanctions become lower than foregone GDP growth. Karmann and Maltritz (2002) use foreign reserves and net exports as an indicator of the ability to pay, which is assumed to follow an Ito process. Using market quotes of the risk-free rates and the current country spreads, the authors calibrate a Merton-like model using the Black-Scholes formula for a put option.[2] The approach by Claessens and Pennacchi (1996) can be seen as a borderline case as they assume an unobservable variable to represent the debt servicing capacity in a first passage time model. Debt capacity is modeled to follow a random walk with drift where default occurs when the process hits zero. By using a Kalman filter, they calibrate their model to Mexican Brady bonds.[3]

4.1.2 Reduced Form Models

Default risk in structural models arises endogenously from the course the repayment capacity takes in relation to the debt obligations. In contrast, reduced form models assume default risk as an exogenously specified process and therefore avoid specifying the payment obligation

---

[2] Their model calibration for four countries, however, shows similar drawbacks as the initial Merton (1974) model in that the long-term risk spreads are much too high.

[3] For other structural approaches to sovereign risk see also the unpublished work of Westphalen (2001) and Rocha and Garcia (2005).

and debt capacity. When this exogenous process concerns the probability of default only, this is referred to as an intensity-based approach. Under risk neutrality, implying that no risk premium compensates for the uncertainty about the realized outcome, the intensity process specifies the instantaneous probability of a credit event. It is therefore also referred to as hazard rate process. While this process remains an abstract, unobservable measure without providing any link to the fundamentals, it proves extremely useful for pricing defaultable claims of any sort, partly due to its computational convenience. A large number of authors have made significant progress in this area, such as Heath et al. (1992), Jarrow and Turnbull (1995), Madan and Unal (1998), Lando (1998), Duffie and Singleton (1999), to mention just a few.

Among the reduced form models, there are two broad subcategories. Equilibrium models follow the methodology of the term structure models for risk-free rates and obey the no-arbitrage rule. In general, arbitrage pricing prevents the possibility that an investor can "lock in" an arbitrage profit by conducting offsetting transactions (such as a credit-financed spot purchase and a future sale). This theoretically appealing feature ensures that the dynamic evolution of yields is consistent with the current term structure of yields, i.e., that time series and cross sectional data are dimensionally consistent. Assuming a similar stochastic process for the hazard rate, the risk adjusted discount rate simply becomes the sum of the risk-free spot and the hazard rate under risk neutrality. Affine term structure models are based on this result and build the discount rate from a linear function of stochastic state variables, assigning each variable a different meaning (such as short and long-term risk premium, or liquidity premium).[4] Another advancement of these models was made when the state variables were allowed to be correlated, referred to as non-linear affine models (Longstaff and Schwartz (1992), Balduzzi et al. (1996), Dai and Singleton (2000), Duffee (2002)). In empirical applications, this characteristic is useful in a deep and efficient bond market which applies to U.S. Treasuries rather than Ecuadoran eurobonds. While the fit of a multi-factor affine model can be satisfactory, it can quickly become very complex and abstract, which is problematic for empirical estimations.[5]

This became an argument to consider parsimonious models for fitting the term structure of risk spreads (Diebold et al. (2005)). While receiving less attention in academic finance, approaches of this kind are popular among market practitioners and central bankers. A first

---

[4] See Duffie and Kan (1996).
[5] See Duffee and Stanton (2004).

subclass are linear functions, in their simplest form consisting of a constant hazard rate which is nothing other than the (mean) bond spread. The exponential polynomial form suggested by Nelson and Siegel (1987) has gained particular notice in this class because it combines intuitively meaningful parameters into a functional form which is flexible enough to fit most term structures and can easily be estimated by non-linear least squares. The basic Nelson-Siegel model allows for any extension of its functional form, such as supplementing another linear factor in the extended Nelson-Siegel model (Svensson (1994, 1995)). Any linear model is in principal convertible to an affine interpretation.[6] A second subclass is constituted from probability distributions at the point in time a credit event occurs. As before, the simplest variant is the constant hazard rate. The resulting likelihood that no credit event occurs until a certain time in the future follows an exponential distribution. Spline approximation techniques, often referred to as non-parametric fitting, can be considered as a third subclass (McCulloch (1971), McCulloch (1975), Vasicek and Fong (1982), and others). Depending on the functional form an arbitrarily close fit can be achieved, resulting in smooth term structure curves of illustrative beauty. All spline methods, however, are sensitive to the selection of knot points. Exponential splines avoid another problem of empirical models, namely unstable or even negative forward rates (Shea (1985)).

The empirical literature of intensity-based models provides an extensive number of studies both with affine and parsimonious models. Until recently, only a few contributions focused on risky sovereign bonds, but this area currently enjoys a significant increase in academic interest. With regard to affine models, Pages (2001) estimates a two process model under the Cox-Ingersoll-Ross square-root specification for the 2024 Brazil Brady bond. While no robustness check (by using the implied parameters to price other equal ranking Brazil bonds) is undertaken, the author compares the implied default risk to stripped Brady yields and the Brazil EMBI spread. The implied term structure of credit spreads displays a characteristic hump shape with steeply increasing spreads over the first two years and a flattening thereafter. Duffie et al. (2003) set up a very rich affine model of the risk spread, considering default, soft restructuring, and political regime switches. Their empirical estimation for Russian MinFins around the Russian crisis 1998, however, uses a much simpler model with a one-dimensional stochastic process for the risk spread. Choosing the MinFin III as benchmark in the model, Duffie et al. (2003) find that it is difficult to explain

---

[6] For a two-factor affine Nelson-Siegel model see Diebold et al. (2005).

the considerable price derivations of other bonds during this turbulent episode. A four factor affine model, with Vasicek-like random walk processes but without allowing for correlation, is implemented by Berardi and Trova (2003). Estimated parameters for six major emerging countries are yielded in a two-step approach. A yield curve is extracted from bond prices by cubic spline fitting, which, in a second step, serves as input for the Kalman filter estimation. The results appear reasonable although the first draft of the working paper does not provide a further analysis of the results. Among parsimonious models, Merrick (2001) assigns a linear function of the default probability and estimates the implied default parameters for Russian and Argentine bonds during the 1998 crisis to search for contagion effects. The linear function applied is, however, not bound between zero and one. Overcoming this drawback, Merrick (2005) uses a simplified two factor Nelson-Siegel model.[7] The resulting parameter estimates are very similar to those by Andritzky (2005) using an extreme value type I probability distribution to model the default rate. The resulting estimates provide an excellent fit to the data and nicely describe the course of the Argentine crisis. Since the extreme value distribution is defined for negative variates as well, a case distinction becomes necessary.[8] This causes an immediate price discount for all bonds, resembling the empirical observation that maturing bonds do not converge to par during a crisis. With regard to the spline method, Izvorski (1998) strips Brady bonds of seven countries and calculates non-parametric densities of the term structure of risk premia, using a normal kernel.

The resulting parameter estimates are often interpreted as describing the default probability. While they might indeed be reasonable proxies for the real default probability, this conclusion can be misleading for two major reasons. First, models often assume risk neutrality and omit other factors of influence. Risk neutrality assumes that an average investor does not require an extra return for bearing risks. In turn, bond spreads only compensate for the expected credit loss, aside from any Jensen inequality term. Although a plausible assumption in finance, it is reasonable to assume that investors of risky bonds are not any different than equity investors, who apparently demand positive and time variant risk premia. Similar excess returns are found in empirical studies of bond spreads (resulting in the "credit spread puzzle") and

---

[7] See Narag (2004) for a working paper with similar methodology.

[8] If the density is larger than zero for negative values of t, the case distinction (introduced to avoid truncation) results in a survival probability of less than one at point $t = 0$. See Andritzky (2005), p. 99.

are interpreted as the result of risk aversion and other effects.[9] One of the effects leading to this wedge might be the misspecification of the risk-free rate (Duffee (1996)) which is mitigated by using swap rates instead of Treasury rates when looking at US dollar bonds. Other effects justifying a risk premium are illiquidity (Longstaff (2004)), difficulties in default risk diversification (Amato and Remolona (2003)), systematic risk and contagion (Cornell and Green (1991), Fama and French (1993), Collin-Dufresne et al. (2003)), and interest rate risk (Collin-Dufresne et al. (2001)). Furthermore, empirically detected risk premia might simply be the result of incorrect (or biased) default expectations (Hull et al. (2005)). While it is possible to calibrate models to distinguish these factors and proxy investors' risk aversion (e.g., Zhang (2003)), it remains problematic to derive real-world default probabilities from the few historical default cases in the sovereign sector. This contrasts with corporate default statistics which offer a sufficient sample size of historical defaults. Given this lack of empirical data, the combined impact of the above effects on sovereign bond spreads is difficult to assess. As any unobservable measure of default risk might be influenced by such effects, it is advisable to keep the caveat in mind until future research sheds more light on these questions.

A second shortcoming requires mentioning. Before calculating default probabilities from risk spreads, an assumption about the recovery value has to be made since it is not given endogenously in reduced form models.[10] As shown in Chap. 3, recovery from sovereign credit events can widely vary. This makes it even harder to interpret the estimated parameters or use them for pricing recovery contingent claims, as for instance credit default swaps in Chap. 6. The existing theoretical and empirical literature, unfortunately, paid little attention to this point.

4.1.3 Recovery Schemes

The following reviews the common concepts of how to model recovery values. Most authors confine themselves to one of the following three schemes which essentially model recovery as a fraction with different denominators:

---

[9] Altman (1989) first pointed out that implied risk-neutral default probabilities are on average lower than realized default rates, suggesting that bondholders earn more than the risk-free rate on average.

[10] The implicit assumption of these models is typically independence of the hazard and the recovery rate. See Duffie and Singleton (1999), p. 688.

Recovery of market value (RMV): The recovery ratio is a fraction of the current market value (Duffie and Singleton (1997), Duffie and Huang (1996), among others).

Recovery of face value (RFV): The recovery ratio is a portion of the nominal value received immediately upon the credit event (Duffie (1999*b*), Duffee (1998), Lando (1998), Brennan and Schwartz (1980)).

Recovery of treasury (RT): The recovery payout is a fraction of the present value of the claim's nominal value, discounted at the risk-free rate. The recovery ratio is therefore a fraction of the price of a risk-free zero bond with similar maturity (Collin-Dufresne and Goldstein (2001), Jarrow and Turnbull (1995), Longstaff and Schwartz (1995)).[11]

Among these schemes, the RMV concept is clearly dominant in academic and applied credit risk modeling, while the other concepts are more cumbersome to implement. The beauty of the RMV scheme lies in its computational tractability since the recovery rate becomes interwoven with the risk premium, making the distinction between hazard and recovery rate effectively irrelevant.

The choice of the recovery approach is both a theoretical as well as empirical issue, asking which concept is actually prevalent in the investors' minds irrespective of its validity or soundness. From a theoretical stand point, it is shown in Chap. 6 that for par instruments (such as credit default swaps) the RFV is conceptually the adequate approach (because after a credit event the face value is reimbursed), while other schemes can be misleading. Chapter 3 suggests a mixed approach to recovery, allowing for equalizing and differentiating recovery values. In an empirical assessment of corporate bond data, Bakshi et al. (2001) show that the RT rather than the RMV approach fits BBB rated bonds best. Regardless of the soundness of different conceptual approaches, most past studies of credit risk in sovereign bonds have avoided this question and applied an RMV scheme.

While the literature on realized sovereign recovery rates is in its infancy, there is ample evidence of physical recovery rates of corporations.[12] On corporations, the literature has so far focused on the

---

[11] The suggestion by Bakshi et al. (2001) of a "recovery of outstanding value" can be classified as a variant of the RT approach. It assumes that the original bond is not replaced by a risk-free zero bond of similar maturity, but rather by a risk-free bond with features similar to the original claim.

[12] Besides publications from private rating agencies see, for example, Altman and Kishore (1996), and Franks and Torous (1994).

statistical distribution of past recovery rates and their determinants, partly utilizing the difference between junior and senior corporate debt (Renault and Scaillet (2004), Unal et al. (2003), Das and Tufano (1996), Jarrow (2001)). Some evidence, while inconsistent, points towards a negative correlation between actual default frequencies and bond recovery rates (Altman et al. (2002), Hu and Perraudin (2002), Frye (2000a,b)). The existence of rich data on past defaults and a bankruptcy-led workout process did not draw much attention on how to calculate expected recovery rates from pre-default bond data. While understandable for corporate debt instruments, the lack of a large quantity of sovereign bond defaults and the constantly evolving design of sovereign workouts pose a different problem. Extracting the recovery value expected by investors from market data is therefore an important and promising field of research. Merrick (2001) and Andritzky (2005) estimate RFV parameters from Argentine and Russian bonds, while Zhang (2003) extracts this parameter from Argentine CDS data. Chapters 5 and 6 continue this string of the literature.

4.1.4 Outlook

As the list of unpublished work on emerging market bonds grows by the day, hope rises that appropriate and accepted models for risky sovereign bonds will soon emerge in academia. Besides the literature on Brady bonds, few authors have stressed addressing the particularities of emerging market sovereign bonds. This attempt will be made in the following, emphasizing the prevalence of soft restructurings instead of outright defaults in sovereign bond markets.

The approach of this study combines reduced form models (which require no other input than bond market prices, which are available as high frequency time series) with the usual virtue of structural models that yield an estimate of the recovery value (but rely on low frequency macroeconomic data). The empirical estimates of a parsimonious model in Chap. 5 result in a series of unobservable variables, which bear a straightforward interpretation as risk neutral default parameters and expected recovery ratio. The results offer insight into the continuous shift of the determinants of sovereign default risk, allowing them to be interpreted by means of macroeconomic fundamentals or economic and political events. This level of frequency of implied default parameters can be achieved with parsimonious models, while richer models, considering a broader range of effects, yield only one set of time-invariant parameters from a sufficient time series of bond data. The approach chosen in this study, on the one hand, bears more restrictions and foregoes the no-arbitrage principle, but, on the other hand, leads to a useful compression of information and performs well in forecasting models.

Using a reduced form model instead of a structural model for yielding estimates of unobservable parameters of sovereign risk avoids the common difficulties when directly linking economic fundamentals to bond prices or spreads. While the link between the bond data and fundamentals is somehow more intuitive in these kinds of models and recovery is often given endogenously, structural models are usually difficult to implement empirically and often prove an unsatisfactory fit to the data. This might also be a result of the relative paucity of data and the evolving and idiosyncratic nature of sovereign risk.

Besides this contribution, the potential for further advancement in the literature on sovereign bond pricing remains considerable. While term structure models for sovereign bonds indicate rapid development, the next generation of the literature could also address portfolio credit risk with sovereign bonds of multiple classes and issuers. This study, however, focuses on researching one class of bonds (defined by pari passu ranking and cross-default clauses) in individual countries, effectively resulting in a perfect correlation of credit risk and recovery within each sample of bonds.

4.2 An Overture to Bond Analysis

This section offers a short introduction to the fundamental concepts of bond pricing and credit risk. The section lays the foundation to understand how modeling credit risk enters into the formulas for the bond price and bond yield.

4.2.1 The Money Market Account and the Discount Factor

The concept of a money market account reflects the idea of a virtually riskless investment which continuously accrues the risk-free rate prevailing in the market. Recognizing that in reality any investment bears some risk, some might prefer the term "reference rate" instead.

Definition 1 (Money market account) $B(t)$ *is defined as the value of a money market account for $t \geq 0$ where, at time $t = 0$, the value of the account is $B(0) = 1$. As the risk-free rate r_t is a positive function of time, the value of the money market account is*

$$B(t) = \exp\left(\int_0^t r_s \, ds\right). \tag{4.1}$$

This can also be written in the form of a differential equation,

$$dB(t) = r_t B(t) \, dt, \text{ with } B(0) = 1. \tag{4.2}$$

The interest rate r_t is the instantaneous rate at which the money market account accrues at the point in time t. This can be written in the more familiar form of a first order expansion,

$$B(t + \Delta t) = (1 + r_t \Delta t) B(t), \tag{4.3}$$

which is similar to Definition 1 for an arbitrarily small time interval $[t, t + \Delta t]$. Throughout this study, notations of interest rates always refer to the instantaneous rate which compounds continuously.

The money market account often serves the reverse question: What is the value at time $t = 0$ of one unit of currency received at time $t = T$? This question is answered by multiplying the amount received with the inverse of $B(T)$, given that the future values of r_t for $t = [0, T]$ are deterministic. In a more general form, this concept of discounting future payments under stochastic risk-free rates is formalized in Definition 2:

Definition 2 (Stochastic discount factor) *The stochastic discount factor $d(t, T)$ equals the amount of currency at time $t \geq 0$ that has the same value as a similar amount of currency received at a time $t = T$ with $t < T$, such that*

$$d(t, T) = \frac{B(t)}{B(T)} = \exp\left(-\int_t^T r_s \, ds\right). \tag{4.4}$$

In the real world, the risk-free rate is neither riskless in the sense of zero volatility, nor guaranteed in the sense of a law of nature. In empirical applications, the risk-free rate needs to be approximated by the interest rate offered by systemically essential institutions within one market. In practice, these are the rates implied from government bonds of the most solvent sovereign issuers in a currency, predominantly the U.S. Treasury bonds for dollar interest rates. Alternatively, short rates charged between banks provide a useful benchmark. This is especially popular in European markets. The London Interbank Offered Rate (LIBOR) is the most frequently used rate, while similar measures exist for other markets as well (such as the EURIBOR, the Brussels counterpart). A term structure curve of zero rates is derived either from stripped yields of government bonds, or from interbank short and forward rates. Since these measures exist only for discrete points in time, rates need to be interpolated. The resulting curve can be used directly in empirical models. Alternatively, no-arbitrage models of the short rate are estimated from those discrete risk-free rates, such as the Cox et al. (1985a) model.

4.2.2 The Price of a Risky Zero Bond

The price of a risk-free zero bond paying one unit of currency at maturity is already given by the logic of the money market account, because such a bond's price equals the applicable discount factor. In case of default risk, reduced form models of credit risk assume the existence of an exogenous default process. Let us assume that this is a risk-neutral hazard rate process λ which determines the default time T'. This can be interpreted as binary process Λ which takes the value 0 before the credit event and 1 thereafter. Under a given Martingale process $\mathbb{Q}$, the risk-neutral hazard rate process λ can be thought of as the jump arrival intensity of a Poisson process.[13] Λ can therefore be written as

$$d\Lambda_t = (1 - \Lambda_t)\,\lambda_t\,dt + dM_t \,, \qquad (4.5)$$

where M is a Martingale under $\mathbb{Q}$. If, in case of a credit event, the holder of a zero bond is compensated by some (reduced) amount of φ, the present value at time t of a risky zero bond paying one unit of currency at time T becomes

$$Z(t,T) = d(t,T)\,(1 - \Lambda_t) + \int_t^T d(t,s)\,\varphi\,d\Lambda_s \,. \qquad (4.6)$$

This combines to a continuous pricing formula for a risky zero coupon bond (Duffie and Singleton (1999), pp. 696f):

Definition 3 (Risky zero bond price) *If the risk-neutral hazard rate process λ is defined under an equivalent Martingale measure $\mathbb{Q}$, a risky zero bond paying one unit of currency at maturity T has the present value at time t of*

$$Z(t,T) = \exp(-\int_t^T r_s + \lambda_s\,ds) + \int_t^T \exp(-\int_t^s r_u + \lambda_u\,du)\,\varphi\,\lambda_s\,ds \,, \tag{4.7}$$

where r_t is the risk-free rate and φ is a deterministic amount of currency paid right upon the arrival of a credit event.

---

[13] A Martingale is a sequence of random variables, parameterized by time, whose expected future value, conditional on the past, is its current value. A Poisson process is a particular type of random process including discontinuities ("jumps").

This setting assumes that there is just one kind of credit event, typically associated with default. In case of sovereigns the typical credit event to be modeled this way would be a restructuring rather than a payment default. However, both events are usually treated similarly in credit risk models, using the term "default" for simplicity.

4.2.3 The Price of a Risky Coupon Bond

The price of a coupon bond can be combined from zero bond prices which resemble the promised cash flows of a coupon bond. The following defines the price of such a fixed coupon bond, with bullet amortization, subject to credit risk.

Definition 4 (Risky coupon bond price) *Let the risk-neutral hazard rate process be* λ, C_t *the time-continuous coupon payment at the instant time* t, *and* N *the bond's notional value repayable at time of maturity* T. *The resulting price of a risky coupon bond is*

$$
P_c(t,T) = \int_t^T \exp(-\int_t^s r_u + \lambda_u \, du) \, C_s \, ds + \qquad (4.8)
$$
$$
\exp(-\int_t^T r_s + \lambda_s \, ds) \, N +
$$
$$
\int_t^T \exp(-\int_t^s r_u + \lambda_u \, du) \, \varphi \lambda_s \, ds ,
$$

where, as before, r_t *is the risk-free rate and* φ *the deterministic recovery amount.*

The assumption of coupons being paid continuously results in what is called the "clean" bond price. This price does not consider accrued but not yet due interest when coupons are paid periodically. Furthermore, the formula needs adaption for special features common in risky sovereign bond contracts such as capitalizing coupons, step-up coupons, or amortization.[14] This is easily achieved when converting Equation 4.8 into a setting with discrete payments:

Definition 5 (Dirty risky coupon bond price) *Let the continuously compounding risk-free rate be* r_t *and* λ *be a risk-neutral hazard rate*

---

[14] Capitalizing coupons are not paid in cash but added to the bond's face value. Step-up (step-down) coupons follow a contractual pattern of increasing (decreasing) coupon payments, possibly linked to some conditions such as GDP growth or inflation. Amortizing bonds start repaying portions of their notional value (often together with coupons) before their final maturity.

process. C_i are $i = 1, ..., I$ coupon payments made at times t_i. N_i are amortization payments coinciding with coupon payments after g grace periods where maturity occurs with the last payment made, $T = t_I$. At t_0, the price of a risky coupon bond, including accrued interest, is accordingly

$$P(t_0, T) = \sum_{i=1}^{I} C_i \exp(-\int_{t_0}^{t_i} r_u + \lambda_u \, du) \qquad (4.9)$$

$$+ \sum_{i=1+g}^{I} N_i \exp(-\int_{t_0}^{t_i} r_u + \lambda_u \, du)$$

$$+ \varphi \sum_{i=1}^{I} \exp(-\int_{t_0}^{t_i} r_u du)$$

$$\left(\exp(-\int_{t_0}^{t_{i-1}} \lambda_u du) - \exp(-\int_{t_0}^{t_i} \lambda_u du) \right).$$

Note that all payments made, C_j, N_i, and φ, are in units of currency. Capitalizing coupons result in a zero payout while adding to the bond's face value so that the sum of all amortization payments may exceed the initial notional.

It could be argued that not only coupons, but also default should realistically be assumed to occur on discrete points in time because insolvency usually becomes obvious upon the debtor's failure to make a due payment. However, this might not be an appropriate assumption for sovereigns for two reasons. First, debtor countries may have several issues with cross-default clauses outstanding, resulting in several payment dates per year. Second, neither the declaration of a debt moratorium, nor the initiation of a restructuring offer have typically occurred exactly on payment dates.

Bond prices are usually quoted as "clean" prices without accrued interest. Information on accrued interest is reported separately by data providers. Since it is assumed to be risk-free, accrued interest can be easily computed under the following standard day count conventions:

Definition 6 (Day count conventions) *When calculating the days of accrued interest, different conventions apply as to what fraction of an annual (or analogously semi-annual or quarterly) coupon this accounts for. There are four conventions:*

Actual/actual. *The year fraction is calculated from the actual days of accrued interest divided by the number of days in the particular year, including leap years.*

Actual/365. *A year always has 365 days while the time span between two dates* T_1 *and* T_2 *is the actual number of days between them, which is denoted by* $T_2 - T_1$. *The year fraction is therefore*

$$\frac{T_2 - T_1}{365}.$$

Actual/360. *As before, but a year consists of only 360 days, therefore the year fraction becomes*

$$\frac{T_2 - T_1}{360}.$$

30/360. *This convention also uses 360 days per year while each month is assumed to have 30 days. If dates are comprised by day, month, and year, that is* $T_1 = (d_1, m_1, y_1)$ *and* $T_2 = (d_2, m_2, y_2)$ *respectively, the year fraction is given as follows:*

$$\frac{\max(30 - d_1, 0) + \min(d_2, 30) + 360(y_2 - y_1) + 30(m_2 - m_1 - 1)}{360}.$$

The convention used is typically defined in the bond prospectus although certain market places and bond contracts traditionally use a certain day count convention. The 30/360 definition is common in the U.S. for fixed rate contracts. Actual/360 is mostly used in the money market and for floaters. Actual/actual is common in the euro area and Great Britain. To avoid distortions for single trading days which can emanate from day count conventions, this study always uses actual/actual.[15]

4.2.4 Yields, Spot and Forward Rates

When considering yield, this study usually refers to the continuous yield to maturity which is the discount rate that causes the present value of all payments to match the observed bond price.

Definition 7 (Yield) *The yield is a constant which serves as the continuously compounded discount rate at which any instrument's present value equals its current trading value.*

---

[15] For example, a distortion may emerge on March 1st when a 30/360 convention is used while the hazard process is time-continuous. Because the observed dirty bond price jumps up on March 1st, the estimated default parameters could show a downward bias. The opposite effect may prevail on the 31st day of a month.

For the dirty value of a risky coupon bond as in Definition 5, the corresponding yield y, given a trading price of the dirty bond value, needs to be numerically determined from

$$P(t_0, T) = \sum_{i=1}^{I} C_i \exp(-y(t_i - t_0)) + \qquad (4.10)$$

$$\sum_{i=1+g}^{I} N_i \exp(-y(t_i - t_0)).$$

The yield can be calculated for any single bond, while many equally ranking bonds of the same issuer can be used to calculate a yield curve analogously to an interest rate term curve. The continuously compounded yield y can always be converted into an annual yield Y by

$$Y = \exp(y) - 1. \qquad (4.11)$$

The (risk) spread, $s = y - r$, refers to the difference between the yield and the risk-free interest rate and can also be calculated on both a time-continuous or an annual base. When using a term structure of the risk-free benchmark rate, the corresponding maturity has to be chosen appropriately to the expected lifetime horizon of the yield.

Above bond valuation formulas already use the instantaneous rate r_t which is the risk-free spot rate at time t. Similarly, the hazard rate λ_t, the yield y_t, and the spread s_t can also be thought of as spot rates. In the same way a term structure of the risk-free rate is given for different time horizons, other measures can likewise be combined into a term structure for different time horizons. Such a term structure can then be used to calculate forward rates, for instance the spot interest rate in one year from today.

Definition 8 (Forward rate) *A forward rate, denoted by $F_0(t_i, T_i)$, is the implied rate at time $t = 0$ for the future time span from t_i to T_i. Given a term structure curve of the spot rate r_t, the implied forward rate $F_0(t_i, T_i)$ is calculated by*

$$F_0(t_i, T_i) = \frac{\int_{t_i}^{T_i} r_u \, du}{T_i - t_i}. \qquad (4.12)$$

The knowledge of one term structure (either the term structure of interest rates, spot rates, or forward rates) is sufficient to infer the others. The yield can be inferred from the forward curve as well since it is an equally weighted average of the instantaneous forward rates.

4.2.5 Default Probability Functions

To describe the likelihood of a credit event to happen, credit risk analysis draws on the standard theory of probabilities. The following distinguishes three kinds of characteristic functions which describe the probability structure of a credit event. Following standard convention, the term "default" is used in these models primarily to signify any kind of credit event.

The survival function is the probability that a credit event does not occur before a certain point in time. It is monotonically declining and bound to the interval $[0, 1]$.

Definition 9 (Survival function) *The survival function $S(t)$ is the probability that the event time, τ, occurs after than any point in time, t:*

$$S(t) = \text{Prob}[\tau > t] = 1 - Q(t), \qquad (4.13)$$

where $Q(t)$ is the cumulative distribution function of the credit event time.

The function $q(x)$ is the density function of the default probability so that $Q(x) = \int_{-\infty}^{x} q(s)\, ds$. In most credit default frameworks t is the time of default as seen from time zero, so that t is always positive. The unconditional default function has therefore the property of $\int_{0}^{\infty} q(s)\, ds = 1$, and stays within the interval $[0, 1]$.

Definition 10 (Unconditional default function) *The unconditional default function, also called unconditional failure rate, is the density of the default probability at any point in time t:*

$$q(t) = \text{Prob}[\tau = t] = \lim_{\Delta t \to 0} \left(S(t) - S(t + \Delta t) \right). \qquad (4.14)$$

The hazard function is the actuarial term for the conditional default density function. There are a number of synonyms for the hazard rate such as conditional default probability, instantaneous failure rate, instantaneous forward rate of default, or simply default intensity.

Definition 11 (Hazard function) *The hazard rate, denoted by λ, is the probability of default at any point in time t, given no default up to that time. The hazard function $\lambda(t)$ is derived from the Bayes' theorem of conditional probability:*

$$\lambda(t) = \lim_{\Delta t \to 0} \frac{\text{Prob}[t < \tau \leq t + \Delta t \mid \tau > t]}{\Delta t} = \frac{q(t)}{S(t)}. \qquad (4.15)$$

The hazard function can also be derived from the survival function since

$$\frac{\partial \ln S(t)}{\partial t} = \frac{1}{S(t)} \frac{\partial S(t)}{\partial t} = -\frac{q(t)}{S(t)},$$

so that

$$\lambda(t) = -\frac{\partial \ln S(t)}{\partial t}. \tag{4.16}$$

This connects the variables introduced in Definitions 9 to 11.

4.2.6 Bootstrap Analysis

The bootstrap analysis is a powerful standard tool to analyze a cross section of equally ranking bonds under the assumption of a simple "recovery of market value" definition.[16] The illustrative description in the following helps to connect Sects. 4.2.4 and 4.2.5 and serves as an example of how to use information contained in a cross section of defaultable bonds.

Let $B^{-1}(t,T)$ denote the price at time t of a risk-free zero bond maturing at time T. Analogously, $Z(t,T)$ is the price at time t of a T-maturity defaultable bond. The continuously compounded yield to maturity of the risk-free bond, y^B, and risky bond prior to default, y^Z, satisfy

$$y^B(t,T) = -\frac{\ln B^{-1}(t,T)}{T-t} \tag{4.17}$$

and

$$y^Z(t,T) = -\frac{\ln Z(t,T)}{T-t}. \tag{4.18}$$

These formulas allow the determination of the implied zero rates for each maturity in the simple case of a cross section of zero bond prices with different maturities. For n bonds maturing after $T_1, ..., T_n$ periods, these n zero rates determine the bootstrapped zero curve. Usually, the zero rate is treated as piecewise constant to gain a stair function of the implied zero curve. In many practical applications, for instance to price a bond issue of new maturity, spline approximations are used to smooth the zero curve for maturities between the bootstrapped zero rates.

Similar to formulas from Sect. 4.2.4, a bootstrapped zero curve from riskless and defaultable bonds can be used to infer the term structure of spreads and forward rates. Prior to default, the credit spread $s(t,T)$

---

[16] See Sect. 4.1.3.

is simply the yield difference of the defaultable and risk-free bond of the same maturity T,

$$s(t, T) = y^Z(t, T) - y^B(t, T). \qquad (4.19)$$

Let us assume that there is only default risk and investors are risk neutral. Then the credit spread has to compensate for the probability of default, adjusted by the residual recovery value, thereby building a bridge to Sect. 4.2.5:

$$s(t, T) = -\frac{\ln(1 - (1 - \psi)Q(T))}{T - t}, \qquad (4.20)$$

where ψ is the recovery rate and $Q(T)$ is the cumulative distribution function of default between time t and T. Note that only under the assumption of zero recovery (i.e., $\psi = 0$) the latter equation transforms to

$$s(t, T) = -\frac{\ln S(T)}{T - t}. \qquad (4.21)$$

The forward interest rates can be calculated by using the term structure of zero rates implied by the cross section of risk-free bonds. If the continuously compounded zero rate for the period from 0 to t_1 is denoted by y_1 and the corresponding periodical rate from 0 to t_2 is denoted by y_2, then the forward rate from t_1 to t_2 is calculated analogous to Definition 8, by

$$F_0(t_1, t_2) = \frac{y_2 t_2 - y_1 t_1}{t_2 - t_1}. \qquad (4.22)$$

The instantaneous forward rate at time t for indefinitely small time intervals is gained from (4.22) by looking at two neighboring points on the term structure curve and taking the limit,

$$f(t) = \lim_{t_1 \to t_2} \frac{y_2 t_2 - y_1 t_1}{t_2 - t_1}. \qquad (4.23)$$

By combining this with (4.17) and (4.18), we can define the instantaneous forward rate from risk-free bonds,

$$f^B(T) = -\frac{\partial \ln B^{-1}(T)}{\partial T}, \qquad (4.24)$$

and from defaultable zero bonds prior to default,

$$f^Z(T) = -\frac{\partial \ln Z(T)}{\partial T}. \qquad (4.25)$$

Analogously to the credit spread $s(t, T)$, an instantaneous forward credit spread is defined as the difference between $f^Z(T)$ and $f^B(T)$. Combining (4.24) and (4.25), as well as (4.17) to (4.21), we see that

$$f^Z(T) - f^B(T) = -\frac{\partial \ln(1 - (1 - \psi)Q(T))}{\partial T}. \qquad (4.26)$$

This is the recovery-adjusted hazard rate. If ψ is set to zero, this becomes more obvious since (4.26) simplifies to

$$-\frac{\partial \ln(1 - Q(T))}{\partial T} = -\frac{\partial \ln S(T)}{\partial T} = \lambda(T). \qquad (4.27)$$

Therefore, this so called instantaneous forward credit spread is nothing but the hazard rate from (4.16), possibly adjusted by the inclusion of some recovery value ψ.

The bootstrapping analysis can also be applied to coupon bonds. After sorting bonds by their maturity date, bootstrapping yields a stair term structure of forward rates. When deducting the corresponding risk-free interest rates, the resulting curve shows the term structure of forward risk spreads. Figure 4.1 shows an example of implied forward credit spreads for sovereign bonds of Argentina, Mexico, and Turkey.

Although handy, the bootstrapping analysis has a number of drawbacks. First, there will always be only as many zero rates determinable as there are quoted instruments with different maturities. The rest of the implied zero curve has to be approximated by some interpolation scheme. The resulting shape of the curve does, to a certain extent, depend on the interpolation method used which does not necessarily obey the no-arbitrage criterium. As bootstrapping determines one point of the curve from one bond, there is no remaining degree of freedom that allows the calibration of other credit-risk parameters, such as the recovery value. Second, in empirical estimations the method is vulnerable to outliers. Since the zero rates are gained through an iterative process beginning with the bond that has the shortest maturity, all of the following implied zero rates $y(t, T_{j+1})$ are a function of their j precedents, $y(t, T_1), ..., y(t, T_j)$. As a consequence, one single mispriced bond will result in distorted rates for all subsequent maturities. Frequently, the resulting forward rates become negative or unrealistically large.

4.2.7 Bond Duration and Average Life

Besides time to maturity, two other measures are often mentioned to describe the length of time in a bond contract. The duration, often

Fig. 4.1. Implied forward risk spreads

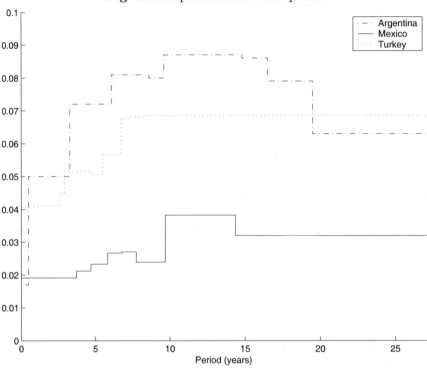

Source: Datastream, author's calculations. Empirical forward risk spreads of three countries calculated by bootstrapping analysis. The Argentine term curve corresponds to the settlement date August 22, 2000, at which all bonds were rated BB by Standard & Poor's. The Mexican curve stems from May 6, 2002, when all bonds were rated BBB-. The Turkish B- rated bonds were taken from January 22, 2002.

referred to as Macauley duration, is a standard measure for this purpose and is calculated as the present value weighted mean repayment time using the risk-free discount factor.

Definition 12 (Duration) *Given a coupon bond with time-continuous coupon payments C_t and bullet amortization of the notional N at maturity T, its duration is given by*

$$D_t = \frac{\int_t^T s \exp(-\int_t^s r_u \, du) \, C_s \, ds + T \exp(-\int_t^T r_s \, ds) \, N}{\int_t^T \exp(-\int_t^s r_u \, du) \, C_s \, ds + \exp(-\int_t^T r_s \, ds) \, N}. \tag{4.28}$$

The average life of a bond is the mean repayment time of the notional and has already been used in the previous chapter. In contrast to the duration formula, the payments do not get discounted.

Definition 13 (Average life) *A bond which amortizes by $j = 1, ..., J$ payments N_j at times t_j has an average life of*

$$L_t = \frac{\sum_{j=1}^{J} t_j F_j}{\sum_{j=1}^{J} F_j} . \tag{4.29}$$

4.3 Functional Forms of the Term Structure

Much research effort has been devoted to term structure modeling with the objective of finding the best way to describe yield, forward, or spread curves. The purpose of this effort is manifold. Economic models support general economic intuition and link the observed term structure to macroeconomic fundamentals, allowing one to forecast interest rates (Campbell and Shiller (1991)), inflation (Fama (1975), Mishkin (1990)), or economic activity (Chen (1991)). Analogously, finance models serve purposes such as asset pricing, hedging, and portfolio allocation. Section 4.1 has already referred to the distinction between equilibrium and empirical models.

All of these approaches have advantages and disadvantages. The following properties present a selection of useful criteria for the evaluation of term structure models:

Consistency of rates. Yields, spreads as well as their forward rates need to be consistent, which normally means non-negativity and continuity as well as some smoothness. Most models though do not comply in all regards.

Convergence. The short rate should lead to $Z(t, T)$ converging to the face value for $t \to T$, while the long rate should converge to some reasonable, non-negative value. While the first condition is necessary for equilibrium models, the latter bestows plausibility upon empirical models.

No-arbitrage. Consistency with regard to the closing of arbitrage opportunities is an essential property for asset pricing but is less relevant in empirical curve fitting, macroeconomic assessments, and term structure forecasting.

Flexibility. The model needs to allow for flexibility to fit a wide range of theoretically possible and empirically observed curves. Although flexibility is always desirable, a trade-off exists between computational tractability (especially when using arbitrage-free asset pricing techniques) and empirical fit.

Further desirable properties from an econometric standpoint are the tractability of the model as well as its suit for efficient estimation methods, for instance by exploiting the time series dependence of the data.

While this list is never complete, it gives a good lead for the following assessment of term structure models in the context of risky sovereign bonds. The scope is limited to affine, polynomial models and probability models as they best serve the purpose of extracting estimates of implied default and recovery parameters.

4.3.1 Affine Models

Affine models use stochastic state processes under a risk-neutral probability measure, mostly to model the bond yield. The state process can be a multidimensional Markov-Ito-process,

$$dX(t) = \mu(X(t), t)\, dt + \sigma(X(t), t)\, dW(t) , \tag{4.30}$$

where $\mu(X(t), t)$ is the drift function and $\sigma(X(t), t)$ is a covariance matrix between the increments of a multidimensional Wiener process, $dW(t)$. The most popular version is a basic one-dimensional square-root process for the risk-free rate as in Cox et al. (1985a) which prevents negative values for r:

$$dr(t) = k(x - r(t))\, dt + \sigma\sqrt{r(t)}\, dW(t) , \tag{4.31}$$

where k, x and σ are positive constants. Defining the short rate as a differential equation complies with the no-arbitrage argument because the forward rates are solutions to the equation. The resulting price of an otherwise risk-free zero bond $Z(t, T)$ can be shown to be

$$Z(t, T) = A(t, T)\, \exp(-B(t, T)\, r(t)) , \tag{4.32}$$

where

$$A(t, T) = \left(\frac{2\gamma \exp((k + \gamma)\,(T - t)/2)}{(\gamma + k)\,(\exp(\gamma\,(T - t)) - 1) + 2\gamma}\right)^{2kx\sigma^{-2}} \tag{4.33}$$

and

$$B(t, T) = \frac{2\,(\exp(\gamma\,(T - t)) - 1)}{(\gamma + k)\,(\exp(\gamma\,(T - t)) - 1) + 2\gamma} \tag{4.34}$$

with $\gamma = \sqrt{k^2 + 2\sigma^2}$. This closed form can be used for coupon bonds as well and obeys the boundary condition $Z(T, T) = 1$.

Affine models combine stochastic processes of several state variables, adding up to the discount rate. Brennan et al. (1979), for example, use two processes to describe the short-term and long-term interest rate. Their setting avoids the one-factor model's disadvantage of perfect correlation between the short-term and the long-term rates. Alternatively, different processes can be used to separately model the reference curve and the risk spread, so that

$$R_t = \sum_{i_r=1}^{I_r} r_{i,t} + \sum_{i_\lambda=1}^{I_\lambda} \lambda_{i,t} , \qquad (4.35)$$

where the risk-free curve is modeled from I_r stochastic processes, and the hazard rate is the sum of I_λ stochastic processes. Another way to combine the stochastic processes is to model two factors which drive the short-term interest rate and its volatility, as suggested by Longstaff and Schwartz (1992).

The seminal contribution of Duffie and Singleton (1999) has shown that under the recovery of market value presumption, the recovery term (i.e., the last summand in (4.8)) can be subsumed into the risk-adjusted discount rate. The recovery adjusted discount rate for the risk-neutral hazard rate process of default, λ, becomes

$$R_t = r_t + \lambda_t(1 - \psi) , \qquad (4.36)$$

where ψ is the exogenous expected recovery fraction of market value. This makes empirical applications of multi-factor affine models much more tractable as estimations can be based on observed yields instead of observed bond prices. This, however, is not the case for alternative recovery assumptions—especially recovery of face value—for which affine bond price models become computationally cumbersome.[17]

4.3.2 Parsimonious Models

Since the bond analysis in Chap. 5 pursues an empirical purpose, the remaining elaborates on empirical models introducing functional forms of the hazard rate in dependence of time.

In contrast to affine models, the state variables in parsimonious models are assumed to be deterministic. This results in ready estimates of the parameter vector for each observation. The estimation relies on the assumption that each bond price contains some deviation from the model bond price. How estimates are gained by using some distributional assumption for these errors is described in Sect. 4.5.

---

[17] See Bakshi et al. (2001).

The risk-free benchmark curve is not subject to the following models but is assumed to be given, either by empirical observations of risk-free proxy rates (such as U.S. Treasury or swap rates) or fitted values from an appropriately chosen model of the risk-free rate (such as Vasicek (1977), or Cox et al. (1985a)). It has therefore to be borne in mind that the following describes only term structures of the hazard rates. Since spreads are not exclusively a function of the hazard rate, but of the recovery value as well, this does not directly translate into a definable term structure of spreads.

In the following, the suggested functional forms for modeling the hazard rate will be illustrated according to Definitions 9 and 11.

Constant Hazard Rate Functions

The simplest case is a constant default probability during a certain period. In a time continuous setting, this is described by a constant instantaneous probability:

Hazard function 1 (Constant hazard rate) *In a constant hazard rate model, a constant variable $\lambda \geq 0$ describes the instant probability of a credit event given that no credit event occurred before, so that*

$$
\begin{aligned}
\lambda(t) &= \lambda\,, \\
S(t) &= e^{-\lambda t}\,.
\end{aligned}
\tag{4.37}
$$

Under zero recovery, the hazard rate corresponds to the forward spread under continuous compounding. The resulting discount factor declines monotonically, fulfilling the properties of no-arbitrage and consistency. Convergence is assured as well, while the model's flexibility is limited. Note that the survival function of the constant hazard rate model follows an exponential distribution. Figure 4.2 illustrates the resulting term structures along the Definitions 9 and 11.

Nelson and Siegel (1987)

The starting point of the Nelson-Siegel model is a second-order differential equation to describe monotonic, humped or S-shaped forward rate curves. Allowing for unequal roots results in an over-parametrization of this model.[18] Restricting the differential equation to unique roots

---

[18] This occurs when the roots become equal and can be avoided, for instance, by a corresponding penalty term added to the likelihood function. See Söderlind and Svensson (1997), p. 419.

Fig. 4.2. The constant hazard rate model

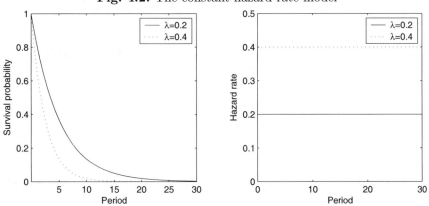

Source: Author's calculations. Constant hazard rate default functions with two different values for λ during a 30 period time horizon.

makes the model more parsimonious while allowing for the same degree of flexibility in fitting term structure curves. The model can be used to fit both risk-free and risk-prone term structures. Analogously, it can be used to describe the course of the hazard rate.

Hazard function 2 (Nelson-Siegel-Model.) *The Nelson and Siegel (1987) polynomial consists of the long-term rate $\beta_0 > 0$, a short-term component β_1, and a medium term component β_2 where $\gamma > 0$ is the (common) decay parameter of the exponential decay terms:*

$$
\begin{aligned}
\lambda(t) &= \beta_0 + \beta_1 \exp(-t/\gamma) + \beta_2(t/\gamma)\exp(-t/\gamma)\,,\\
S(t) &= \exp\Big[-\beta_0 t - \beta_1 t\frac{1-\exp(-t/\gamma)}{t/\gamma}\\
&\quad -\beta_2 t\Big(\frac{1-\exp(-t/\gamma)}{t/\gamma} - \exp(-t/\gamma)\Big)\Big]\,.
\end{aligned}
\tag{4.38}
$$

The short-term hazard rate for $t = 0$ is therefore $\beta_0 + \beta_1$ while in the long-term it converges to β_0. The short-term factor β_1 can also be considered as an indicator of the slope of the hazard curve. The hump-shape (U-shape) of the hazard rate curve is determined by $\beta_2 > 0$ ($\beta_2 < 0$) which has its maximum impact at $t = \gamma$.

The original Nelson and Siegel (1987) model was extended by Svensson (1994, 1995) by allowing for another hump- (or U-) shape. This is easily done by adding another summand $\beta_3(t/\gamma_2)\exp(-t/\gamma_2)$ to the hazard rate function. This allows modeling of even more complicated hazard rate structures while adding more parameters to the model. Any number of similar extensions is imaginable. To avoid over-parametrization these extensions are omitted here.

Fig. 4.3. The Nelson and Siegel (1987) model

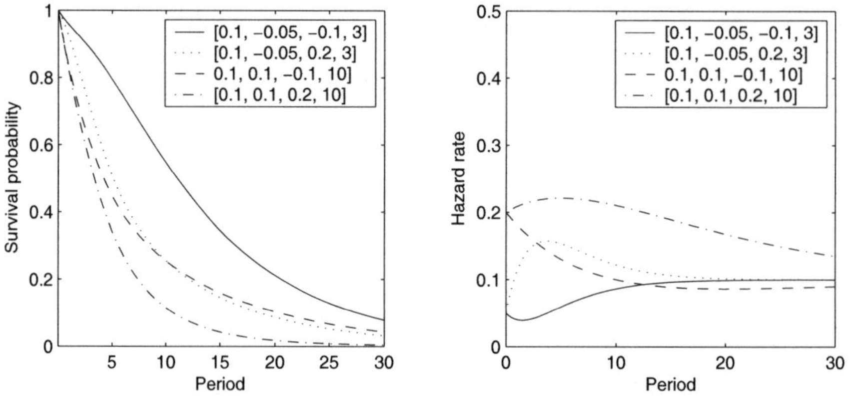

Source: Author's calculations. Survival and hazard function derived from the Nelson and Siegel (1987) model for different sets of $\Theta = [\beta_0, \beta_1, \beta_2, \gamma]$ during a 30 period time horizon.

The Weibull Distribution

Another set of hazard functions can be derived from standard continuous univariate probability distributions, using them to model the unconditional default probability. The advantage of this approach is that these functions obey the standard boundary conditions and are sufficiently researched. Analogous approaches are widespread in other disciplines such as engineering where product reliability has to be assessed.

Since the time variable is usually defined to be positive, a suitable probability density function should assign non-zero probabilities only to positive values of t. Otherwise, a case differentiation becomes necessary to ensure the survival probability to be one at the starting time $t_0 = 0$.

The extreme value theory suggests that the Weibull distribution correctly models the failure time when many competing failure processes are combined. An example from sports helps explaining: In a race of a large number of equivalent competitive participants, the winner's time is likely to follow a Weibull distribution. In an economic analogy, the Weibull distribution should successfully model the default time when the default of a sovereign debtor is determined by risk factors of roughly equal probability such as political disruptions, external shocks, etc.

The Weibull distribution is determined by a shape and a scale parameter. A third parameter of the Weibull distribution, the location parameter, is set to zero here so that the distribution is only defined for non-negative values of t. The Weibull hazard function can show constant ($\gamma = 1$), decreasing ($\gamma < 1$) and increasing ($\gamma > 1$) intensities (see Figure 4.4). The scale parameter stretches or compresses the run of the curve.

Hazard function 3 (The Weibull distribution) *Besides time* t, *the two-parameter Weibull distribution (also called extreme value distribution type III) is determined by the shape parameter* $\gamma > 0$, *and the scale parameter* $\alpha > 0$:

$$\begin{aligned}\lambda(t) &= \tfrac{\gamma}{\alpha}\left(\tfrac{t}{\alpha}\right)^{\gamma-1}, \\ S(t) &= \exp\left(-(t/\alpha)^{\gamma}\right).\end{aligned} \tag{4.39}$$

Fig. 4.4. The Weibull distribution

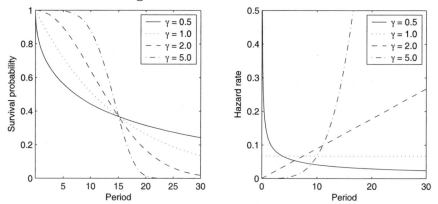

Source: Author's calculations. Survival and hazard function derived from the Weibull distribution for different values of γ, while $\alpha = 15$ during a 30 period time horizon.

For $0 < \gamma \leq 1$, the probability density function is monotonically declining with a non-existent mode. In Figure 4.4, this results in a sharply decreasing and convex slope of the survival function. The hazard rate is convex and monotonically declining.[19] For $\gamma = 1$, the Weibull density becomes a special case of the exponential distribution with a flat hazard rate. For $\gamma > 1$, the hazard rate is monotonically increasing.[20] For $1 < \gamma < 2$ the curve is concave and turns convex for $\gamma > 2$. The resulting survival function shows an inflection point.

---

[19] In engineering, this type of failure rate is associated with so called "early-type failures" of just manufactured products, for which the failure rate decreases with age.

[20] This correspond to "wear-out types of failures" in engineering.

The scale parameter α can be thought of as having the same units as t and stretches the density function since it is directly proportional to the function's standard deviation.

The properties of probability distributions guarantee smooth and consistent rates. Since the two-parameter Weibull distribution is only defined for positive values of t, it ensures convergence of maturing issues to their face value. For $\gamma > 1$, however, the hazard function is monotonically increasing, leading to unbounded long-term rates, although this drawback does not necessarily create problems within the usual range of maturities.

The Gumbel Distribution

The disadvantage of the Weibull distribution is that the shape parameter γ creates very distinctive hazard rates making results difficult to interpret. This reason has made the Gumbel maximum distribution—another extreme value distribution—an attractive alternative. Mathematically convenient to handle and easy to interpret, this distribution depends on two parameters.

Hazard function 4 (The Gumbel distribution) *The Gumbel distribution (also called extreme value distribution type I) is defined by the location parameter α (which is the mode of the distribution density) and the shape parameter β (which is proportional to the density's standard deviation),*

$$\lambda(t) = \frac{1}{\beta}\frac{\exp\left(-(t-\alpha)/\beta\right)}{\exp\left(\exp\left(-(t-\alpha)/\beta\right)\right)-1},$$
$$S(t) = 1 - \exp\left(-\exp\left(-(t-\alpha)/\beta\right)\right). \tag{4.40}$$

Note that the distribution can assign positive values to the density when t is negative. This caveat must be addressed by a case distinction for the unconditional default function in discrete time between t and $t + \Delta$:

$$Q_{t+\Delta} - Q_t = \begin{cases} e^{-e^{-\frac{\Delta-\alpha}{\beta}}} & \text{for } t = 0 \\ e^{-e^{-\frac{t+\Delta-\alpha}{\beta}}} - e^{-e^{-\frac{t-\alpha}{\beta}}} & \text{for } t > 0 \end{cases} \tag{4.41}$$

Given values of α and β at which the distribution is positive for values of t below zero, this case differentiation results in an immediate non-zero unconditional default probability.[21] Although this violates the

---

[21] For $\alpha = 5$ and $\beta = 3$, the immediate default probability at time $t = 0$ implied by (4.41) would be 16%, for example.

convergence condition, it can be argued that this makes sense empirically. During distress, it can be observed that bonds trade at a discount until maturity. Bondholders are aware that the amortization payment remains at risk until settlement because default often occurs upon maturity if re-financing was not achieved.[22]

Since the mode of this density is α, the Gumbel distribution allows modeling an unconditional probability curve which shows a maximum at time $t = \alpha$ and has a standard deviation of $\frac{\pi}{\sqrt{6}}\beta$. Each parameter determines a different characteristic of the function. For instance, a change of α only leads to a change in the location of the distribution function without affecting the shape of the function. This makes this distribution very illustrative.

Fig. 4.5. The Gumbel distribution

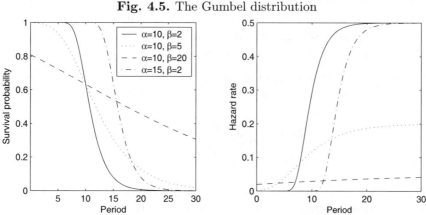

Source: Author's calculations. Survival and hazard function of the Gumbel distribution for different combinations of α and β during a 30 period time horizon.

The density of the Gumbel distribution has a constant skewness of 1.14 and kurtosis of 5.4. The hazard rate converges to the inverse of the shape parameter β which avoids unrealistically large long run rates under most circumstances. The typical term structure curves this distribution forms are U-shaped due to the initial discount induced by the case distinction and the monotonically increasing hazard function. The model's flexibility is therefore limited.

---

[22] This was actually the case in the Dominican Republic where the restructuring offer was launched during the grace period. However, unless default is imminent, maturing bonds trade at lower yields rather than higher yields compared to the average of all bonds.

The Lognormal Distribution

Because observed forward spreads are not a monotonically increasing (or decreasing) function, but show an initial increase in the medium term and a decrease for long maturities, the lognormal distribution might be an attractive alternative. Together with the Weibull distribution, the lognormal distribution is a very popular assumption in reliability applications. In this regard the model can be motivated by applying the central limit theorem to small multiplicative shocks contributing to default. This is called "multiplicative degradation" and has been used in semiconductor failure models. Assuming that default occurs when this product reaches a critical amount, the default time is described by a lognormal distribution. In the context of sovereign default, it sounds economically intuitive that a credit event is triggered by the multiplicative effect of shocks, such as disruptions in trade, a speculative attack on the exchange rate, or a bank run.

In a simplified setting (assuming a zero location parameter), the distribution is described by two parameters.

Hazard function 5 (The lognormal distribution) *The function is defined by a shape parameter, $\beta > 0$, and a scale parameter, $\tau > 0$:*

$$
\begin{aligned}
\lambda(t) &= \frac{1}{\beta t} \frac{\theta\big((\ln t - \ln \tau)/\beta\big)}{\Theta\big(-(\ln t - \ln \tau)/\beta\big)} \, , \\
S(t) &= 1 - \Theta\big(\tfrac{\ln t - \ln \tau}{\beta}\big) \, ,
\end{aligned}
\tag{4.42}
$$

where Θ is the cumulative normal distribution function and θ is the probability density function of the normal distribution.

In this model, the time to default assumes a lognormal distribution. This implies that the natural logarithm of the default time follows a normal distribution with mean τ and standard deviation β. The lognormal distribution is skewed to the right, positively related to both the scale and shape parameters. This model obeys the convergence criterium since t must remain positive so that $S(t_0) = 1$.[23] The hazard function can build a monotonically downward sloping curve or a hump shape, but shows a limited flexibility otherwise (see Fig. 4.6). A drawback for advanced calculations might be the lack of a closed form solution for the cumulative density.

---

[23] This is at least true for the above version of the lognormal distribution which does not consider a location parameter.

Fig. 4.6. The lognormal distribution

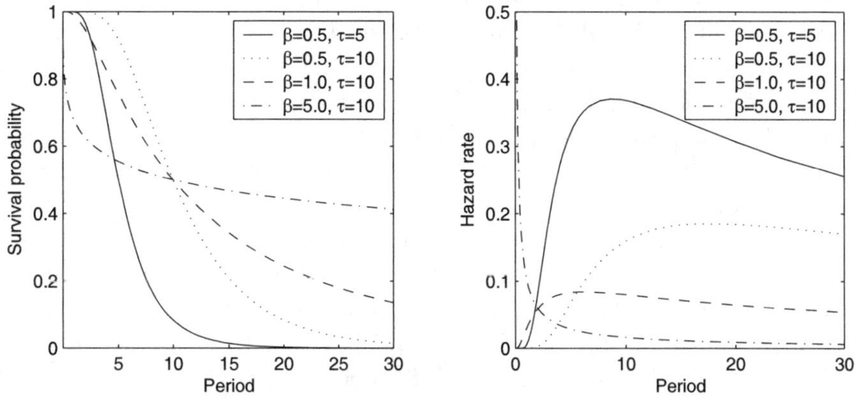

Source: Author's calculations. Survival and hazard function derived from the lognormal distribution for combinations of β and τ during a 30 period time horizon.

4.3.3 Discussion

The discussion of the above models leads back to the dilemma that all approaches have some advantages and disadvantages. Affine models provide a sufficient fit only when using multiple stochastic factors. As affine models satisfy all of the evaluation criteria outlined in the beginning of this section they have been at the forefront of empirical analysis of credit spreads. For this study, however, there are two obstacles. Firstly, as of today, affine models cannot accommodate more sophisticated schemes of modeling recovery without resorting to very burdensome computations. Secondly, affine models are not designed for estimating default parameters cross-sectionally. This, however, is the objective of the empirical section in Chap. 5 which estimates implied default parameters to be used as feed for fundamental models of sovereign risk.

Among parsimonious models (the route followed in this study), the most popular model by Nelson and Siegel (1987) presents the benchmark other models are tested against in the first part of Chap. 5. However, the Nelson-Siegel model suffers from some serious drawbacks. Even in its standard form, the model results in parameter estimates inconsistent with common properties of term structures since there are essentially no restrictions which ensure positivity and finiteness of all rates. To avoid problems on the short end of the curve, a handy solution is to set the short rate equal to the hazard rate of the shortest bond. This, however, is only makeshift, leaving aside many other parameter combinations which can lead to inconsistent rates. Extremely small values of the decay term can indicate this. On the long end of the curve, a combination of an unrealistically large value of the long rate can go hand in hand with a large decay rate. This leads to a good fit only within the maturity range of the sample since the long rate slowly converges to β_0. To conclude, the Nelson-Siegel model is a powerful empirical tool, although its ample flexibility can make some adjustments necessary in practice.

The beauty of distributional models certainly is that they satisfy the consistency and smoothness criteria. Furthermore, they fulfill the convergence criterium (except for the Gumbel distribution), and are well established in similar applications of related disciplines. Among the distributional models, the Weibull model is easy to implement and follows the appealing intuition of failure rate models. It has been used extensively in related applications and provides reasonable flexibility. However, the parameters are not as straightforward to interpret as, for instance, in the Gumbel model. The Gumbel model proves to be a capable model under distress-like circumstances (for instance in the case study of the Argentine crisis in Andritzky (2005)) as it does not require prices to converge to par at maturity. However, the model lacks widespread use and presents only limited flexibility as the hazard rate function is always positively sloped. Finally, the lognormal distribution resembles economic intuition by modeling the multiplicative effect of adverse shocks. One disadvantage is, however, that the shape parameter α can be difficult to interpret. Even though the model seems to be mathematically manageable, one has to bear in mind that there is no closed formula for the cumulative distribution function of the standard normal distribution.

Finally, the constant hazard rate model is certainly the most simple model, following the plain assumption that a constant fraction of the surviving claims defaults each period. While complying with most criteria, foremost no-arbitrage, the model lacks any flexibility in fitting non-constant term structures of hazard rates. However, in combination with a recovery of face value scheme, a constant hazard rate model forms inverted term structures of bond spreads in times of distress while resulting in a flat term structure when there is no distress. The constant hazard rate model is therefore applied in Chap. 6 where the empirical setup requires the most parsimonious model without compromising on the no-arbitrage condition.

For the sake of the empirical analysis of bonds in Chap. 5, distributional models will be compared with the Nelson and Siegel (1987) model using data from Brazil. The empirical performance will supplement the above discussion as to which model to pick for the subsequent application on half a dozen sovereign emerging market issuers. The hazard rate models chosen thereby allow for incorporating alternative concepts of modeling recovery. The following section is dedicated to developing such a model, reflecting the nature of sovereign default and restructurings.

4.4 Modeling Recovery

Modeling recovery can be viewed from three perspectives. Chapter 3 has analyzed the recovery value which emerged from recent restructurings. This ex post view helps to establish certain patterns of how recovery is constituted. Besides distinguishing soft and hard restructurings, the review showed the typical features of restructuring offers such as maturity extension, treatment of past due interest, and nominal value write offs. An analysis of this sort is, however, only a snapshot of the past. The pattern of recovery develops in lock-step with the changing nature of sovereign restructurings. Lately, Argentina set a new milestone by its harsh restructuring, and other countries might follow suit. The resulting variety makes it desirable to model recovery values with as much flexibility as possible. This suggests a multifaceted approach to modeling recovery which could include separate recovery fractions for the nominal value and coupons, modeling maturity extension and coupon reduction, and considering delays in the workout process.

The above view appears incompatible with the perspective of the academic literature which is dominated by the recovery of market value (RMV) scheme (see Sect. 4.1.3). The RMV model looks at recovery in a very restrictive and narrow way. While the RMV model proves easy to implement in intensity based models of credit risk, it does not allow any conclusion regarding the expected recovery implied in traded instruments. Alternative frameworks, such as the recovery of face value (RFV) remain a burdensome way of modeling stochastic recovery rates.[24] Section 3.4.2 of the previous chapter described the realized recovery values from recent sovereign restructurings with the RMV and RFV frameworks. For the hard restructuring in Argentina, the outcome suggests a 38% RFV fraction while the RMV fraction is insignificant. The soft restructuring of Uruguay resulted in a 53% RFV fraction and suggests a considerable RMV fraction. This finding is supported by the theoretical thoughts in Sect. 3.4.2. It has been argued

---

[24] See Bakshi et al. (2001).

that pari passu ranking claims upon acceleration should result in recovery values independent from their pre-default value when creditor equity is enforced. Furthermore, intuition as well as empirical observations confirm the existence of a lower bound of the bond value even when the credit event is inevitable. Both arguments contradict modeling recovery as a fraction of the pre-default market value only.

The following adopts a third perspective, carefully striking a balance between the variety of ways the recovery values can actually be comprised and the information content provided by sovereign bond prices. On one hand, it is not a sensible proposition that all market participants assume detailed expectations on all facets of a future restructuring. It is an impossible endeavor to extract market expectations on haircuts, coupon reductions, or maturity extension from bond prices. On the other hand, bond prices should contain some information about investors' expectations on recovery values which goes beyond the abstract concept of bond spreads. As a compromise, the following promotes the concept of mixed recovery, combining RMV and RFV schemes in one model and resembling the distinction of soft and hard restructurings.

The remainder of this section proceeds as follows. The classic RMV scheme is described first and serves as a benchmark which is compared to a mixed approach. The mixed recovery scheme does not rule out that some fraction of the recovery is aligned with the pre-default market value, but allows a recovery fraction which is proportional to the outstanding par value. This approach is called mixed recovery and presents a more careful interpretation of the empirical credit risk parameters than previous approaches such as Merrick (2001, 2005), Duffie et al. (2003), and others.

4.4.1 Recovery of Market Value

This concept is predominant in the credit risk literature and enjoys high popularity among practitioners. The simplicity induced by the RMV scheme may have contributed to a widespread perception of irrelevance of recovery for credit risk analysis. To recall the underlying assumption, the following provides a formal definition of the RMV recovery scheme.

Recovery scheme 1 (Recovery of market value) *In this concept, the value received upon a credit event is assumed to be*

$$\varphi = \psi P(t_-, T), \tag{4.43}$$

where $\psi \in [0, 1]$ is the recovery fraction and $P(t_-, T) = \lim_{t \to \tau} P(t, T)$ is the bond's price one instant before the credit event.

As previously mentioned, the RMV recovery term can be subsumed into the risk-adjusted discount rate for an exogenously expected recovery rate ψ, as in (4.36). The resulting risk spread is the product of the intensity of default, λ, and ψ. In turn, the RMV scheme results in an identification problem when both λ and ψ are assumed to be endogenous. As a consequence, the RMV concept describes the default loss risk in one handy figure—the credit spread—which incorporates the event's intensity and the corresponding recovery rate.

4.4.2 Mixed Recovery

The concept of mixed recovery introduced in this study is an expansion of the traditional recovery of face (RFV) scheme. While it allows the recovery to be proportional to the initial claim's face value it does not rule out that a part of the recovery value is proportional to the pre-default market value:

Recovery scheme 2 (Mixed recovery) *The mixed recovery value is assumed to be comprised by two parts,*

$$\varphi = \omega N(t_-) + \psi P(t_-, T), \tag{4.44}$$

where $\omega, \psi > 0$ are the recovery fractions of the face value, $N(t_-)$, and the market value, $P(t_-, T)$, respectively, an instant before the credit event, t_-.

Note that a positive RFV fraction ω provides some kind of a lower bound to the bond value. If the risk-free rate is zero and default intensities are positive, the price of a zero bond converges with longer maturity to this boundary.

The model provides flexibility in so far as for $\omega = 0$ it presents a pure RMV model while for $\psi = 0$ it comprises a pure RFV framework. In an empirical application, however, this enables the estimation of ω under most conditions. In case of bonds trading homogenously close to par so that $N(t) \simeq P(t, T)$, it is obvious that RMV and RFV fractions have the same effect.[25] The resulting identification problem in the RFV model does not allow a precise estimation of ω. Estimates of ω typically show spikes and a large positive correlation to the default intensity.

The motivation for this kind of recovery model is twofold. From an empirical stand point, Chap. 3 has provided some evidence that a model like this is a useful description of actual recovery outcomes. Furthermore, it is simple enough to yield estimates from non-defaulted

---

[25] See Duffie and Singleton (1999), pp. 703ff.

bond prices even before a distress episode. During distress, the default-contingent value can make up a considerable portion of the total bond value, allowing a more detailed analysis of investors' recovery expectations. When bonds trade at lower spreads, however, it might not be possible to extract much information about recovery expectations from bond prices as these nuances are obscured by the usual noise of market prices.

4.4.3 Discussion

Since the term structure of default intensities under RMV looks similar to a down-scaled term structure of credit spreads, their shape look very familiar. Under distress, the term structure is typically inverted with declining spreads for longer maturities. The left-hand plot in Fig. 4.7 illustrates this for single trading days before, during, and after the Brazil crisis in 2002. Some recovery of face fraction, however, provides a lower bound for bond prices, especially for longer term issues. The resulting term structure of default intensities, in turn, remains flat or is slightly increasing (see right-hand plot in Fig. 4.7).

Fig. 4.7. Term structure of intensities for Brazil under RMV and mixed recovery with $\psi = 0\%$ and $\omega = 25\%$

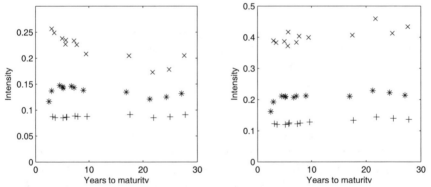

Source: Datastream, author's calculations. Intensity term structure implied from Brazil US dollar bonds on May 22, 2002 (+), July 29, 2002 (x), and January 28, 2003 (*) for a pure RMV model with zero recovery (left plot) and a fixed recovery of face value of 25% (right plot).

A model allowing for an endogenous recovery fraction of face does not therefore necessarily require a high flexibility with regard to term structure curvatures. Simpler, parsimonious models do an equally good job in fitting intensity structures emanating from models which allow for endogenous RFV. This difference, however, is only relevant under considerable default risk while remaining irrelevant when spreads are low.

Empirical estimations of the RFV fraction can lead to an identification problem for high quality issuers. Any model of recovery which allows for more flexible recovery schemes is vulnerable with respect to over-parametrization. Duffie et al. (2003) encounter this problem when distinguishing recoveries from default, restructuring, and regime switches.[26] Even if this specification would take the model of recovery closer to reality, the empirical estimation of all implied recovery parameters is beyond the information content offered by bond price data.

For this reason, the model used in the empirical section of Chap. 5 foregoes further extensions which could take the recovery model even closer to the true model. Among these extensions is the joint stochastic modeling of recovery and default intensity, reflecting the fact that both parameters are correlated and fluctuate over time. Furthermore, the finding in Chap. 3 of systematically positive rents in restructuring transactions suggests the existence of a risk premium on restructurings. This is in line with market sources that indicate that investors prefer to stay away from the restructuring process and sell their position in advance. These extensions outside the scope of this study remain challenges for future research.

4.5 Empirical Implementation

A crucial part of an empirical analysis is the way model parameters are estimated from observed data. Empirical research in emerging markets often faces challenges when it comes to empirical estimations of established models as the data quality is usually inferior to that of industrialized countries. With regard to bond prices, there are only a few countries with sufficient market data to estimate a model as outlined above. Excluding Brady restructurings, many countries have just recently begun to tap the sovereign global bond market (such as Vietnam) or remain unable to expand its issuing activity (such as Ecuador). In the meantime, many of the big market players (along with Russia and South Africa and many transition countries in Eastern Europe) have emerged as investment grade debtors, shedding most of the sovereign

---

[26] See Duffie et al. (2003), p. 128.

risk this study is interested in. This limits the scope of the empirical analysis in Chap. 5 to a few countries for which an empirical analysis can yield insightful results. The following identifies suitable estimation techniques to optimally exploit the bond data of those countries and gain robust results.

An empirical analysis of bond data can be based on two kinds of dependent variables, bond prices and bond yields. Due to the predominance of affine models under the RMV assumption, the empirical literature has relied largely on yield analysis as this is computationally more tractable. Bond prices are analyzed only if required by the underlying model, for instance when using a different recovery assumption.[27] Models of the yield term structure cannot easily be transformed into models of bond prices with definable stochastic properties: if a term structure is, for instance, defined by a Brownian diffusion, model prices will, in turn, be neither homoscedastic nor lognormally distributed for different maturities and coupons.

For recovery assumptions other than RMV, it is necessary to model bond prices rather than bond yields. As there is no simple way to solve a bond price formula for the implied yield, it would be computationally burdensome to base the empirical estimation on fitting a term structure model of yields or spreads.[28] For the model outlined in the previous section, for instance, this would imply a nested two-step optimization.[29] Such an approach proves computationally impossible to tackle.

The empirical implementation is therefore based on cross sections of bond prices. Maximum likelihood is an asymptotically efficient estimation method, even though little is known about the finite-sample properties (especially in the context of the outlined credit risk models). Many applications have switched to quasi maximum likelihood (QML) where the exact probability distribution is unknown or intractable. Only for a few cases, such as the Vasicek (1977) or Cox et al. (1985a) models of the interest rate, there is a closed form solution for the likelihood function, whereas other models need to be solved numerically.

Following the prominence of QML estimations for computational performance, the bond price model is calibrated by minimizing the root mean squared error (RMSE) with regard to the parameter vector Θ,

---

[27] See for example Claessens and Pennacchi (1996), or Merrick (2001).

[28] See Bakshi et al. (2001).

[29] In a first step, the intensity structure has to be inferred for an estimated recovery fraction, which in turn is fitted by the term structure model.

$$\min_{\Theta} \sqrt{\frac{1}{n} \sum_{n=1}^{N} \epsilon_n^{\mathrm{RMSE}}}, \tag{4.45}$$

where

$$\epsilon_n^{\mathrm{RMSE}} = \left(\frac{\hat{P}_n - P_n}{P_n}\right)^2, \tag{4.46}$$

with P_n being the fitted bond model price of bond n in a cross section with N bonds and $\hat{P}_n$ being its observed dirty market price. The RMSE corresponds to the maximum likelihood estimator when bond prices are homogenously distributed.

If yields are homoscedastic (as they are usually assumed to be), bond prices are not homoscedastic. Bond prices with long (short) durations show a high (low) sensitivity to changes in the yield. This is also obvious from the characteristic convergence of bond prices to par at maturity.[30] Bliss (1997) suggests a correction for this heteroscedasticity by weighting the errors with the inverse of the Macauley duration D_n of each bond (see Definition 12):

$$w_n = \frac{1/D_n}{\sum_{n=1}^{N} 1/D_n}, \tag{4.47}$$

which is multiplied with the error term $\epsilon_n^{\mathrm{RMSE}}$ in the minimization function. Comparing estimation results with equal and duration weighting, differences in parameter estimates and in-sample errors prove not to be significant. All estimates are nevertheless conducted with duration weighting.

While this approach has a clear econometric motivation, other weighting proposals are harder to justify. Subramanian (2001), for instance, suggests a corresponding weighting by liquidity (using instrumental variables such as the number of trades, the size of trades, and bid-ask spreads). Less liquid issues, which are presumed to have higher pricing errors, receive less weight in the optimization function.[31] Related corrections could be introduced which control for coupon effects et cetera, but these suggestions remain hypothetical. Chapter 5 will discuss these arguments based on an assessment of the goodness of fit.

Although minimizing pricing errors in the form of QML estimators yields estimates at any point in time, the approach makes no use of the time series characteristic of the data. As a result, the econometrician

---

[30] See, among others, Vasicek and Fong (1982).
[31] This, however, does not address the usual notion that less liquid issues trade at a discount.

has to accept large jumps of the estimated parameters from observation to observation. The course of the parameters over time has a jagged form, sensitive to price movement of single bonds. By defining the unobservable state variables which determine the bond model value as a stochastic process, the time series characteristic of the data could be exploited. However, this requires the additional restriction that the parameter movements assume some stochastic process. The empirical results of this study do not impose this additional assumption as it presents a trade-off in the goodness-of-fit to the weekly data.[32]

4.6 Concluding Remarks

This chapter has illustrated different approaches of default probability models and ways to mathematically describe the recovery value. Introducing the models of hazard rates and recovery in the bond pricing formula, the resulting model is well suited to describe cross-sectional bond prices from sovereigns with substantial default risk and a wide range of maturities and coupons. The following briefly outlines the final model and the estimation strategy of the following chapter, and establishes criteria for model evaluation.

The model estimated in Chap. 5 (and parts of Chap. 6) uses the standard formula for dirty bond prices as in (4.9),

$$
\begin{aligned}
P(t_0, T) = & \sum_{i=1}^{I} C_i \exp\left(- \int_{t_0}^{t_i} r_u + \lambda_u \, du\right) \\
& + \sum_{i=1+g}^{I} N_i \exp\left(- \int_{t_0}^{t_i} r_u + \lambda_u \, du\right) \\
& + \varphi \sum_{i=1}^{I} \exp\left(- \int_{t_0}^{t_i} r_u du\right) \\
& \left(\exp\left(- \int_{t_0}^{t_{i-1}} \lambda_u du\right) - \exp\left(- \int_{t_0}^{t_i} \lambda_u du\right) \right),
\end{aligned}
$$

where the hazard rate λ is modeled along parsimonious hazard functions 1 to 5. Recovery is assumed to follow the mixed concept of (4.44),

---

[32] See Andritzky (2004a) for an example of fitting similar models to sovereign bond data from Russia and Turkey. The implied default parameters are assumed to follow a Gaussian diffusion process. Estimates are gained through an extended Kalman filter. Applying a similar approach to the models used herein resulted in a significantly inferior fit.

$$\varphi = \omega N(t_-) + \psi P(t_-, T).$$

However, the empirical estimations suffer from an identification problem if $\psi > 0$ since this parameter occurs, in a reformulation of the bond price formula (4.9), only in the first two summands as the product $\lambda(1 - \psi)$. Estimations using bond price data in Chap. 5 therefore use the restriction $\psi = 0$ which leads back to a pure RFV scheme. However, both parameters, ω and ψ, are estimated in Chap. 6 when bond and credit default swap data are combined. Implied estimates of the hazard rate parameters and the recovery fraction ω are found by minimizing the root mean squared error from (4.45) and (4.46) using duration weighting along (4.47).

Although all hazard rate models introduced in Sect. 4.3.2 have their virtue, the empirical application will reveal their performance in fitting observed bond prices. The assessment of the model performance requires checking the econometric properties of residuals and testing for the robustness of the estimates. The latter is given when point estimates are not affected by the choice of starting values in the optimization and are robust to small alterations in the sample. These are general prerequisites (to which all models broadly comply), whereas goodness-of-fit criteria play a more decisive role in evaluating the suitability of the model. First, the goodness-of-fit is assessed by the in-sample-fit. In this case, the mean squared residuals of the models are calculated for individual bonds as well as for different subsamples of bonds. The objective is to determine the model with the overall best in-sample fit and discard models that cause biased bond price estimates in sub-samples. Second, the out-of-sample analysis provides another check. Usually, out-of-sample errors are calculated by excluding single observations from the estimation and subsequently, using the estimation result as a forecast for the omitted observation, comparing the forecast to the actual data. In this study, a slightly more sophisticated analysis is employed by implementing a trading strategy based on naive forecasts to test for the predictive ability of the respective model. The trading strategy is based on the assumption that the model indicates the fair price of a bond to which it will converge. In the long-neutral trading strategy, a long position is incurred when the model's price exceeds the market price by at least 1%. Otherwise, the funds are invested at the risk-free rate. The long-short strategy incurs an additional short position when the market price exceeds the model's price by at least 1%. Both strategies are implemented under alternative assumptions of zero trading costs and transaction costs of 25 basis points of the bond's value for each purchase or sale. Besides the returns of such strategies, the Sharpe ratio, i.e. the ratio of excess return to its standard deviation as in Sharpe (1994), is calculated and compared to a simple buy-and-hold strategy of an equally weighted portfolio of the respective bonds. Another out-of-sample indicator is the forecasting ability of the model. The unconditional forecasting probability is the probability that the trading strategy correctly projects the direction of the market. The conditional forecasting probability is the sum of the two conditional probabilities for a correct forecast given market up- and downturns, respectively. According to Merton (1981), this figure is required to be above one as a necessary and sufficient condition for market timing.

5

Empirical Estimations

Based on the models of sovereign bonds in Chap. 4, this chapter presents empirical estimations for major sovereign bond issuers in emerging markets. The analysis serves two purposes. First, it provides some evidence on the adequacy of the bond value models outlined. Indeed, the estimations prove a remarkable in-sample fit of bond value models which allow for the recovery of a fraction of par. Second, implied credit risk parameters for the respective emerging market countries are calculated which allows a cross-country comparison. This advances the analysis of sovereign risk contained in pari passu ranking bonds from simply observing a country's average risk spread.

The data used for this analysis is gathered from mainstream data sources. Information about bond features is drawn from Bloomberg. Time series data of clean mid market prices are obtained from Datastream. To avoid distortions due to differences between bond trading in the U.S. and Europe, estimates utilize data from US dollar denominated bonds only as this is the most common currency of denomination. Some countries with strong relations to the euro area, such as Turkey, have a considerable number of euro denominated bonds outstanding. A few countries have issued bonds denominated in other foreign currencies like the Japanese yen, the British pound, or the Swiss franc. US dollar denominated global bonds therefore offer the largest range of instruments from which to extract information on sovereign risk. Local currency denominated bonds are not the subject to this study as these instruments cannot be compared internationally. The risk-free term structure is directly obtained from US swap rates provided by Datastream.

The remainder of the chapter is structured as follows. In the first section, a comprehensive data set of Brazil bonds is exploited to compare the different models introduced in Chap. 4. Based on the evaluation criteria outlined in the previous chapter, the best performing model is used to obtain estimates for the remaining countries in Sect. 5.2.

5.1 Empirical Model Comparison

Based on the extraordinarily rich data set of Brazil US dollar denominated bonds, the models outlined in Chap. 4 will be estimated and their results compared. The bond sample provides a sufficient data set for the years 2000 through 2004. During this time frame, Brazil was, with close to 20%, the largest weight of the JP Morgan EMBI index family, only temporarily topped by Mexico and Argentina (before 2002). This period reflects a variety of economic circumstances. The country was affected by the Argentine crisis in 2001 and suffered from a serious confidence crisis in 2002. In the last years, the government succeeded in restoring fiscal performance and market confidence, pushing down spreads to historic lows.

In contrast to the EMBIG index which excludes bonds below an outstanding amount of $500 million or of less than six months maturity, the sample applied in this study utilizes all available data. An exception to this are Brady bonds, floating rate notes (the 2004 Brazil 2009 US$ float), and bonds with contractual features difficult to evaluate such as put or call options (the 2000 Bra 2040 US$ 11%). The resulting sample is shown in Table 5.1.

In the following, the alternate hazard rate models from Chap. 4 are compared to a standard Nelson-Siegel model (see Sect. 4.3.2) with the traditional recovery of market value scheme, which serves as benchmark. In contrast to standard applications of term structure models, the models are not fitted to bond yields or spreads, but to the duration weighted bond prices (see Sect. 4.5) to enable different models of endogenous recovery.

5.1.1 The Nelson-Siegel Model

Following the standard approach of term structure estimation, a Nelson-Siegel model as described in Sect. 4.3.2 is estimated, using the standard recovery of market value scheme. In contrast to most other studies, the model is applied on a cross section of bond prices (rather than spreads). Although this takes the error statistics to a different range, the resulting shapes of the term structures are broadly comparable. In accordance with the standard setting of the model, this section assumes a fractional

Table 5.1. Sample of Brazilian global bonds

Name	First cou-pon date	Maturity date	Par ($mln)	Obs	Price range Min	Mean	Max
1997 Bra 2027 10.125% US$	15-Nov-1997	15-May-2027	3500	258	41.51	79.09	114.43
1998 Bra 2008 9.375% US$	07-Oct-1998	07-Apr-2008	1250	258	48.41	86.80	111.85
1999 Bra 2009 14.5% US$	15-Apr-2000	15-Oct-2009	2000	258	57.24	107.35	135.95
2000 Bra 2007 11.25% US$	26-Jan-2001	26-Jul-2007	1500	232	52.25	97.56	118.50
2000 Bra 2030 12.25% US$	06-Sep-2000	06-Mar-2030	1600	252	46.06	92.24	132.25
2000 Bra 2020 12.75% US$	15-Jul-2000	15-Jan-2020	1000	258	47.13	96.04	135.47
2001 Bra 2006 10.25% US$	11-Jul-2001	11-Jan-2006	1500	208	58.78	97.57	112.50
2001 Bra 2024 8.875% US$	15-Apr-2001	15-Apr-2024	2150	198	38.82	72.04	104.06
2001 Bra 2005 9.625% US$	15-Jan-2002	15-Jul-2005	1000	190	63.63	97.26	110.33
2002 Bra 2008 11.5% US$	12-Sep-2002	12-Mar-2008	1250	147	51.63	97.85	119.80
2002 Bra 2012 11% US$	11-Jul-2002	11-Jan-2012	1250	155	45.94	92.49	121.25
2002 Bra 2010 12% US$	15-Oct-2002	15-Apr-2010	1000	142	50.13	97.63	123.88
2003 Bra 2013 10.25% US$	17-Dec-2003	17-Jun-2013	1250	81	87.50	103.43	117.70
2003 Bra 2011 10% US$	07-Feb-2004	07-Aug-2011	1250	74	88.00	103.97	115.88
2003 Bra 2007 10% US$	16-Jul-2003	16-Jan-2007	1000	87	99.70	107.77	114.36
2003 Bra 2019 8.875% US$	14-Apr-2005	14-Oct-2019	833	64	68.08	88.73	103.84
2003 Bra 2010 9.25% US$	22-Apr-2004	22-Oct-2010	1500	63	91.15	103.12	112.00
2004 Bra 2014 10.5% US$	14-Jan-2005	14-Jul-2014	1250	25	97.20	109.64	118.83
2004 Bra 2034 8.25% US$	20-Jul-2004	20-Jan-2034	1500	47	67.25	83.96	97.13

Source: Datastream, Bloomberg, JP Morgan.

recovery of market value in case of a credit event. As suggested in the previous chapter, the bond with the shortest maturity in the sample directly determines the corresponding (short) rate.

Table 5.2 shows the mean of the resulting parameter estimates and their standard deviation. The corresponding term structures are illustrated in Fig. 5.1. They show a characteristically inverted term structure of spreads during the Brazilian crisis in 2002–2003. During this time, the short-term factor β_1 is large and positive. The curvature parameter, β_2, tends to be negative in the earlier periods, indicating a U-shaped term structure, but turns positive later on, resulting in a hump-shaped curvature.

Table 5.3 provides a look at the in-sample accuracy provided by the estimation, illustrated per individual bond and year. Although the data get more complete during the time frame, pricing errors remain broadly homogeneous. While the overall pricing error of 2.07% is much higher than usual model errors from corporate and treasury markets in industrialized countries, the fit is within the same range as comparable studies, such as Merrick (2001). The increase in errors during the time window, peaking in 2002, can be attributed to the higher risk spreads during the Brazil crisis.

Partitioned by year and residual maturity, it becomes evident that the long maturity issues carry the largest pricing error (see Table 5.4). Bonds with a final maturity of less than 10 years, however, do not

Fig. 5.1. Brazilian term structures of the default loss intensity

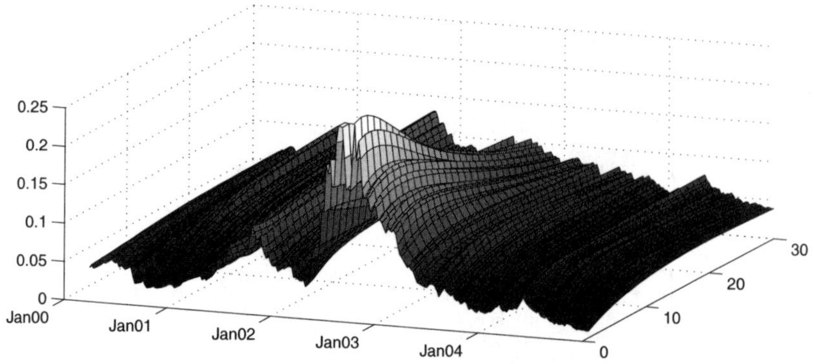

Source: Author's calculations. Surface of weekly default loss intensity term structure implied by weekly estimates of the Nelson-Siegel model for maturities of up to 30 years.

Table 5.2. Estimated weekly parameters for the Nelson-Siegel model

	2000	2001	2002	2003	2004
β_0	0.0820	0.110	0.0555	0.0364	0.0363
	(0.0186)	(0.0135)	(0.0398)	(0.0212)	(0.0203)
β_1	-0.0343	-0.0422	0.0723	0.0143	-0.0241
	(0.0192)	(0.0133)	(0.0895)	(0.0444)	(0.0135)
β_2	-0.0999	-0.0956	0.115	0.134	0.0936
	(0.0697)	(0.0297)	(0.171)	(0.0763)	(0.0505)
$log(\gamma)$	0.506	0.719	1.08	1.74	1.54
	(0.141)	(0.287)	(0.743)	(0.452)	(0.466)
N	43	52	52	52	53
mean(n)	5.5	8.4	11.5	14.0	18.4

Source: Author's calculations. The parameters $\beta_0, \beta_1, \beta_2$ as well as γ determine the curvature of the spread term structure. The table shows mean values of the weekly estimations and their standard deviation for Brazil. N is the number of weekly observations per year while n is the average number of bond prices observed each week.

show a consistent pattern of larger pricing errors for longer maturities. Neither does the direction of the pricing error reveal any regularity.

To indicate the impact of crises on the model's performance, Table 5.5 presents error statistics categorized by individual bond prices and country spreads according to the EMBIG subindex for Brazil. This offers clear evidence that the in-sample fit decreases when spreads rise

Table 5.3. In-sample pricing errors for the Nelson-Siegel model

	2000	2001	2002	2003	2004	Total
1997 Bra 2027 10.125%	2.02	2.22	1.76	1.23	0.965	1.69
1998 Bra 2008 9.375%	2.02	2.89	1.68	0.971	0.527	1.81
1999 Bra 2009 14.5%	2.74	3.08	1.40	1.63	1.10	2.11
2000 Bra 2007 11.25%	1.05	1.04	1.30	0.985	0.566	1.01
2000 Bra 2030 12.25%	0.913	2.05	3.61	2.70	3.31	2.74
2000 Bra 2020 12.75%	1.02	1.79	3.35	2.38	2.55	2.39
2001 Bra 2006 10.25%	-	0.995	1.24	1.79	1.15	1.33
2001 Bra 2024 8.875%	-	2.22	5.00	2.54	1.72	3.17
2001 Bra 2005 9.625%	-	1.60	2.97	3.13	1.30	2.45
2002 Bra 2008 11.5%	-	-	1.23	0.868	0.59	0.908
2002 Bra 2012 11%	-	-	1.60	1.35	0.462	1.23
2002 Bra 2010 12%	-	-	1.23	1.09	0.986	1.09
2003 Bra 2013 10.25%	-	-	-	1.53	0.828	1.12
2003 Bra 2011 10%	-	-	-	0.582	0.434	0.481
2003 Bra 2007 10%	-	-	-	0.725	0.881	0.824
2003 Bra 2019 8.875%	-	-	-	7.15	3.21	4.16
2003 Bra 2010 9.25%	-	-	-	0.568	1.02	0.965
2004 Bra 2014 10.5%	-	-	-	-	0.633	0.633
2004 Bra 2034 8.25%	-	-	-	-	2.09	2.09
Total	1.81	2.12	2.54	1.98	1.56	1.99

Source: Author's calculations. Means of the RMSE in percent for each of the n bonds per year as well as totals.

Table 5.4. RMSE (%) per maturity bucket and year for the Nelson-Siegel model

	2000	2001	2002	2003	2004	Total
0–1 years	-	-	-	-	0.726	0.726
1–5 years	-	1.27	2.22	1.73	0.895	1.52
5–10 years	2.19	2.51	1.38	1.28	0.837	1.55
10+ years	1.41	2.07	3.62	2.75	2.46	2.63

Source: Author's calculations. Means of the RMSE in percent for all N observations of bond prices in each respective category for Brazil.

and prices fall. Again, the direction of the error does not provide evidence of strongly biased estimates.

The out-of-sample analysis (Table 5.6) indicates positive returns for both long-neutral and long-short trading strategies, although these do not outperform a pure buy-and-hold investment, both before and after

Table 5.5. RMSE (%) per EMBIG spread and clean market price for the Nelson-Siegel model

	0–250 bps	250–500 bps	500-1000 bps	1000+ bps	Total
0%–50%	-	-	-	5.22	5.22
50%–75%	-	-	2.15	2.56	2.47
75%–100%	-	2.39	1.85	2.35	1.97
100%–125%	-	1.01	1.72	2.94	1.56
125% or higher	-	1.96	1.24	-	1.83
Total	-	1.38	1.81	2.80	1.99

Source: Author's calculations. Means of the RMSE in percent for all N observations of bond prices in each respective category. Spreads refer to the JP Morgan EMBIG subindex for Brazil.

deducting transaction costs. A good portion of the buy-and-hold return was earned during 2003 and 2004, when Brazil spreads tightened considerably. The active trading strategies maintain overall a smaller exposure to these swings and participate neither in the largest losses nor in the largest gains. The resulting standard deviation is therefore much smaller, resulting in a superior Sharpe ratio for the long-neutral strategy even after trading costs. After costs, the long-short strategy ceases to outperform the passive investment. The unconditional probability of correct forecasts is considerably higher than 50% for the long-neutral strategy only. The second measure of the conditional predictive ability is slightly above the threshold of one, indicating the model does not exhibit a significant predictive ability.

Table 5.6. Performance measures of different trading strategies

Bond	Long-neutral	Long-short	Buy-hold
Compounded return	0.078	0.035	0.13
Simple return	0.081	0.036	0.14
Sharpe ratio	1.2	1.1	0.73
Compounded return after costs	0.071	0.021	0.13
Sharpe ratio after costs	1.1	0.34	0.73
Unconditional probability	0.65	0.52	
Conditional probability	1.1	0.81	

Source: Author's calculations. All returns and Sharpe ratios are based on annualized figures based on trading strategies using price deviations from the Nelson-Siegel model for Brazil.

5.1.2 Two-factor Nelson-Siegel With RFV

Results of the standard Nelson-Siegel model can be compared with a simplified version of the model which restricts β_2 to remain zero. The term structure is therefore solely determined by a short-term and a long-term rate, plus the decay factor γ which describes the term curve's transition from the short to the long rate. To ensure non-negativity of the short rates, $\beta_0 + \beta_1$ is restricted to be equal to or larger than zero, a somewhat weaker restriction than in the standard Nelson-Siegel model. As substitute for the β_2 parameter, the model allows the endogenous determination of a recovery parameter describing the fractional recovery of face value, $\omega \in [0,1]$.[1] While in the previous section the four parameters estimated weekly were $\beta_0, \beta_1, \beta_2$, and γ, this section estimates β_0, β_1, γ, and ω.

Fig. 5.2. Surface of Brazilian intensity term structures

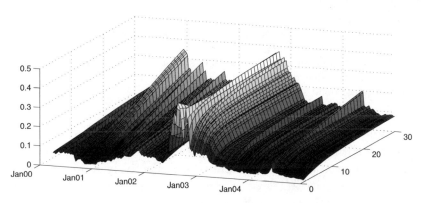

Source: Author's calculations. Intensity term structure implied by weekly estimations of the simplified Nelson-Siegel model with endogenous recovery of face value for maturities between one and 30 years.

This section attempts to evaluate whether an RFV-based model provides a better fit to the bond price data than the previous RMV-based model. This aspect is of particular relevance during periods of crises when the recovery value makes up a substantial portion of the total bond value. While maintaining the same number of parameters to estimate, the result will yield an answer to the question whether a

---

[1] However, it is possible to estimate the standard Nelson-Siegel model together with ω, but results show no significantly improved fit compared to this simplified version.

larger degree of freedom with regard to shapes of the term structure, or a bond model with endogenous recovery describe the bond data better.

Table 5.7. Estimated weekly parameters for the two-factor Nelson-Siegel model with endogenous recovery

	2000	2001	2002	2003	2004
β_0	0.603	0.830	0.344	0.147	0.147
	(0.300)	(0.119)	(0.246)	(0.0431)	(0.209)
β_1	-0.567	-0.775	-0.211	-0.147	-0.142
	(0.322)	(0.114)	(0.311)	(0.0434)	(0.192)
γ	3.80	4.22	2.00	0.166	0.751
	(0.674)	(0.394)	(1.970)	(0.407)	(1.06)
ω	0.297	0.206	0.190	0.286	0.326
	(0.0537)	(0.0646)	(0.0314)	(0.0851)	(0.0533)
N	43	52	52	52	53
mean(n)	5.5	8.4	11.5	14.0	18.4

Source: Author's calculations. The results are the mean and standard deviation of the weekly estimated parameters β_0, β_1 as well as γ, which determine the curvature of the spread term structure. In this version of a two-factor Nelson-Siegel model, β_2 is restricted to remain zero. The model estimates simultaneously the endogenous recovery of face value, ω. N is the number of weekly observations and n the average cross sectional bond sample size.

Table 5.7 displays the resulting parameter estimates and Fig. 5.2 illustrates the resulting shapes of the intensity term structure. Before, Fig. 5.1 showed a term structure which effectively represented the default loss intensity $\lambda(1 - \psi)$ where ψ is the recovery fraction of market value. The resulting recovery value φ equals a fraction of the bond's market price just prior to the credit event, $\psi(t)P(t_-, T)$. Fig. 5.2 displays the corresponding intensity when the recovery value is comprised of an RFV fraction ω. As a consequence of separating the RFV recovery fraction, the intensities in Fig. 5.2 are much higher for given bond prices than in Fig. 5.1.

As indicated in Table 5.7, the RFV fraction ω is significantly different from zero. This means that the RFV model does not collapse into an RMV-style model (which would be the case if $\omega = 0$). Recovery of face value fractions range from 12% to 32%, apparently decreasing during crises. This supports the existence of an additional RMV fraction (in a mixed recovery layout) since the values of ω can be considered too low to constitute the entire recovery value.

However, some caveats apply to the results. The long-term factor β_0 is unrealistically high during the first years. However, this does not come into effect due to the large values of the decay factor γ. It is characteristic for an RFV type model to show flat or increasing intensity curves even when the term structure of spreads is inverted. In 2003, the short-term component β_1 brings the short rate down to zero which is not an unusual result in empirical applications.[2]

Table 5.8. RMSE (%) per bond and year for the two-factor Nelson-Siegel model

	2000	2001	2002	2003	2004	Total
1997 Bra 2027 10.125%	0.649	1.01	0.495	1.01	0.907	0.847
1998 Bra 2008 9.375%	1.14	2.62	1.75	1.13	0.569	1.61
1999 Bra 2009 14.5%	1.58	2.01	1.38	0.798	0.583	1.36
2000 Bra 2007 11.25%	0.901	1.00	1.47	1.24	0.542	1.1
2000 Bra 2030 12.25%	0.661	1.01	1.07	0.744	0.624	0.846
2000 Bra 2020 12.75%	1.06	0.997	1.17	0.819	0.564	0.94
2001 Bra 2006 10.25%	-	0.935	0.967	0.792	0.985	0.923
2001 Bra 2024 8.875%	-	0.872	0.927	1.62	0.865	1.13
2001 Bra 2005 9.625%	-	2.06	1.59	1.59	1.11	1.57
2002 Bra 2008 11.5%	-	-	1.09	0.959	0.555	0.882
2002 Bra 2012 11%	-	-	1.55	1.29	0.496	1.19
2002 Bra 2010 12%	-	-	0.988	0.72	0.801	0.827
2003 Bra 2013 10.25%	-	-	-	1.48	0.728	1.05
2003 Bra 2011 10%	-	-	-	0.587	0.456	0.497
2003 Bra 2007 10%	-	-	-	0.555	0.759	0.686
2003 Bra 2019 8.875%	-	-	-	4.54	2.26	2.79
2003 Bra 2010 9.25%	-	-	-	0.574	0.666	0.652
2004 Bra 2014 10.5%	-	-	-	-	0.348	0.348
2004 Bra 2034 8.25%	-	-	-	-	0.949	0.949
Total	1.06	1.51	1.26	1.21	0.884	1.16

Source: Author's calculations. Mean values of the RMSE in percent for each of the n bonds per year as well as totals.

Results in Table 5.8 provide a better in-sample fit of the model with endogenous recovery of face value than a standard Nelson-Siegel model, although the number of estimated parameters remains the same. The fit improves considerably during the turbulent period between 2001 and

---

[2] With regard to sovereign bond markets, see Alper et al. (2004) for similar findings from Nelson-Siegel models as well as Rocha and Garcia (2005), and Yeh and Lin (2003) for affine models.

2003. Additionally, the overall fit for each bond issue is consistently better, reducing the large errors emanating from the 2003 Bra 2019 8.875% bond. A major portion of the improved fit stems from long-term bonds as the pricing error appears to be less dependent on residual maturity (see Table 5.9).

Table 5.10 illustrates the model's strong performance especially during periods of distress. In comparison to the standard Nelson-Siegel model, the RMSE results are cut in half for spreads above 1,000 basis points, and the error for bonds trading below 50% dropped from 5.2 to 1.2. Although RMSE results are still slightly higher when spreads are high, errors now appear to be unrelated to the bonds' price range.

Table 5.9. RMSE (%) per maturity bucket and year for the two-factor Nelson-Siegel model

	2000	2001	2002	2003	2004	Total
0–1 years	-	-	-	-	0.576	0.576
1–5 years	-	1.48	1.52	1.08	0.787	1.12
5–10 years	1.29	1.99	1.33	1.07	0.625	1.23
10+ years	0.812	0.979	0.950	1.48	1.18	1.14

Source: Author's calculations. Means of the RMSE in percent for all N observations of bond prices in each respective category for Brazil.

Table 5.10. RMSE (%) per EMBIG spread and clean market price for the two-factor Nelson-Siegel model

	0–250 bps	250–500 bps	500-1000 bps	1000+ bps	Total
0%–50%	-	-	-	1.22	1.22
50%–75%	-	-	0.881	1.54	1.41
75%–100%	-	1.47	1.26	1.44	1.31
100%–125%	-	0.682	1.01	1.27	0.929
125% or higher	-	0.389	0.593	-	0.439
Total	-	0.778	1.13	1.48	1.16

Source: Author's calculations. Means of the RMSE in percent for all N observations of bond prices in each respective category. Spreads refer to the JP Morgan EMBIG subindex for Brazil.

Out-of-sample results, however, do not indicate much improvement in predictive ability over the standard Nelson-Siegel model (see Ta-

Table 5.11. Performance measures of different trading strategies

Bond	Long-neutral	Long-short	Buy-hold
Compounded return	0.059	0.044	0.13
Simple return	0.061	0.045	0.14
Sharpe ratio	1.1	1.8	0.74
Compounded return after costs	0.050	0.028	0.13
Sharpe ratio after costs	0.91	0.81	0.74
Unconditional probability	0.67	0.53	
Conditional probability	1.1	0.92	

Source: Author's calculations. All returns and Sharpe ratios are based on annualized figures based on trading strategies using price deviations from the two-factor Nelson-Siegel model for Brazil.

ble 5.11). Returns prove lower for the long-neutral strategy while being slightly higher for the long-short strategy. However, the respective standard deviations appear to be disproportionally lower, resulting in better Sharpe ratios for both active strategies before and after costs. The measures for the predictive ability of the model are slightly better than in the standard Nelson-Siegel model.

5.1.3 The Weibull Model With RFV

This and the following models use functional forms of probability distributions to fit the term structure. While presenting an even more parsimonious parametrization than the previous models, the approach shows a still acceptable in-sample fit. In the following, the Weibull model (see Sect. 4.3.2) is combined with the endogenous estimation of the recovery fraction of face value. The estimated parameters are the shape parameter γ, the scale parameter α as well as the recovery fraction ω.

The estimated time series of implied credit risk parameters excludes seven observations in 2000 which indicated values for γ above 2.2. Such values result in extremely high hazard rates at the long end which exceed one and would be out of the range of the illustration in Fig. 5.3. However, this adjustment does not change the total mean pricing residual. The total number of weekly estimates in 2000 remains similar to those of the previous models as the smaller number of parameters facilitates the estimation with a smaller cross section. While the intensities at the short end are comparable to the simplified Nelson-Siegel model, the Weibull model results in higher long-term intensities, espe-

Fig. 5.3. Surface of Brazilian intensity term structures

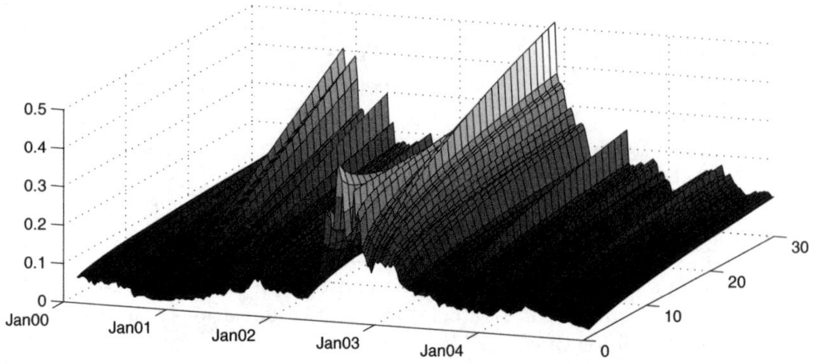

Source: Author's calculations. Term structure implied by weekly estimates of the Weibull model for maturities of up to 30 years.

cially in December 2000 and during all of 2002. However, the overall result closely resembles Fig. 5.2 from the previous model.

Table 5.12. Estimated weekly parameters for the Weibull model

	2000	2001	2002	2003	2004
γ	1.41	1.44	1.14	1.26	1.31
	(0.265)	(0.200)	(0.161)	(0.0894)	(0.0867)
α	11.2	9.24	6.79	8.25	11.7
	(1.12)	(0.735)	(3.16)	(1.76)	(1.64)
ω	0.298	0.193	0.187	0.323	0.364
	(0.0557)	(0.0711)	(0.04)	(0.0807)	(0.0537)
N	42	52	52	52	53
mean(n)	5.2	8.4	11.5	14.0	18.4

Source: Author's calculations. The parameter γ is the shape parameter and α the scale parameter which determine the curvature of the intensity term structure. The table shows mean values of the weekly estimations and their standard deviation for Brazil. N is the number of weekly observations per year while n is the average number of bond prices observed each week.

Table 5.12 displays the means of the resulting parameter estimates. The recovery of face fraction ω is within the same range as in the previous section and shows a high correlation of 97%. The shape parameter γ is above one at all times (except in summer 2002, around the height of the Brazil crisis), indicating an increasing term structure of intensities.

Residual mean squared errors of the bond prices on average show higher values than for the simplified Nelson-Siegel model with endogenous recovery, but are lower than in the standard Nelson-Siegel model (see Table 5.13). This relation holds for both the cross sectional and the time series dimensions.

Table 5.13. In-sample pricing errors for the Weibull model

	2000	2001	2002	2003	2004	Total
1997 Bra 2027 10.125%	0.653	1.02	0.495	0.891	0.976	0.839
1998 Bra 2008 9.375%	1.13	2.94	1.73	1.30	0.658	1.75
1999 Bra 2009 14.5%	1.36	1.89	1.46	0.984	0.668	1.33
2000 Bra 2007 11.25%	0.633	1.06	1.69	1.36	0.986	1.27
2000 Bra 2030 12.25%	0.626	1.16	1.11	1.19	1.43	1.16
2000 Bra 2020 12.75%	0.946	0.971	1.17	0.948	0.55	0.937
2001 Bra 2006 10.25%	-	0.965	0.912	1.44	1.78	1.33
2001 Bra 2024 8.875%	-	0.797	1.00	1.74	0.794	1.17
2001 Bra 2005 9.625%	-	2.47	1.90	2.67	1.79	2.21
2002 Bra 2008 11.5%	-	-	1.27	1.19	0.617	1.05
2002 Bra 2012 11%	-	-	1.70	1.03	0.943	1.26
2002 Bra 2010 12%	-	-	1.02	1.07	1.06	1.05
2003 Bra 2013 10.25%	-	-	-	1.07	0.967	1.00
2003 Bra 2011 10%	-	-	-	0.931	0.853	0.876
2003 Bra 2007 10%	-	-	-	0.614	1.41	1.16
2003 Bra 2019 8.875%	-	-	-	4.39	2.24	2.73
2003 Bra 2010 9.25%	-	-	-	0.644	0.633	0.634
2004 Bra 2014 10.5%	-	-	-	-	0.671	0.671
2004 Bra 2034 8.25%	-	-	-	-	1.02	1.02
Total	0.969	1.62	1.36	1.44	1.16	1.33

Source: Author's calculations. Means of the RMSE in percent for each of the n bonds per year as well as totals.

Similar to the simplified Nelson-Siegel model, Table 5.14 does not reveal higher pricing errors for shorter maturities as the standard Nelson-Siegel model did. On average, however, the Weibull model does not perform as well for shorter maturities as the simplified Nelson-Siegel model. Neither the size of the error nor its direction are conspicuous though.

The categorization of residual errors into EMBIG spread and price buckets shows a slightly weaker performance of the Weibull model in comparison to the simplified Nelson-Siegel model. Nevertheless, the in-sample fit is still very good during distress (see Table 5.15). It appears

Table 5.14. RMSE (%) per maturity bucket and year for the Weibull model

	2000	2001	2002	2003	2004	Total
0–1 years	-	-	-	-	1.26	1.26
1–5 years	-	1.72	1.73	1.56	1.28	1.50
5–10 years	1.17	2.11	1.41	1.15	0.868	1.33
10+ years	0.761	1.01	0.981	1.56	1.29	1.20

Source: Author's calculations. Means of the RMSE in percent for all N observations of bond prices in each respective category for Brazil.

though that the model cannot handle a wide range of bond prices well during distress. In the 1000+ basis points spread category, the RMSE increases with higher bond prices, resulting in a disappointing fit for bonds trading above par in crises. While bonds trading at a low price tend to be overvalued by the Weibull model, high trading bonds tend to exhibit an undervaluation.

Table 5.15. RMSE (%) per EMBIG spread and clean market price for the Weibull model

	0–250 bps	250–500 bps	500-1000 bps	1000+ bps	Total
0%–50%	-	-	-	1.29	1.29
50%–75%	-	-	0.928	1.64	1.50
75%–100%	-	1.49	1.31	1.82	1.42
100%–125%	-	1.04	1.27	2.16	1.21
125% or higher	-	0.648	0.726	-	0.665
Total	-	1.06	1.27	1.68	1.33

Source: Author's calculations. Means of the RMSE in percent for all N observations of bond prices in each respective category. The price category refers to the observed market price without accrued interest. Spreads refer to the JP Morgan EMBIG subindex for Brazil.

The naive trading strategy—performed as out-of-sample test—does not beat a buy-and-hold investment in absolute returns. However, the investment realizes very high Sharpe ratios before as well as after trading costs—the highest among the models considered. The Merton (1981) measures of market-timing show weak evidence of a forecasting ability for the long-neutral strategy and does not significantly differ from the previous models.

Table 5.16. Performance measures of different trading strategies

Bond	Long-neutral	Long-short	Buy-hold
Compounded return	0.073	0.048	0.13
Simple return	0.076	0.049	0.14
Sharpe ratio	1.4	2.2	0.74
Compounded return after costs	0.065	0.031	0.13
Sharpe ratio after costs	1.2	1.2	0.74
Unconditional probability	0.70	0.53	
Conditional probability	1.1	0.82	

Source: Author's calculations. All returns and Sharpe ratios are based on annualized figures based on trading strategies using price deviations from the Weibull model for Brazil.

5.1.4 The Gumbel Model With RFV

Fig. 5.4. Surface of Brazilian intensity term structures

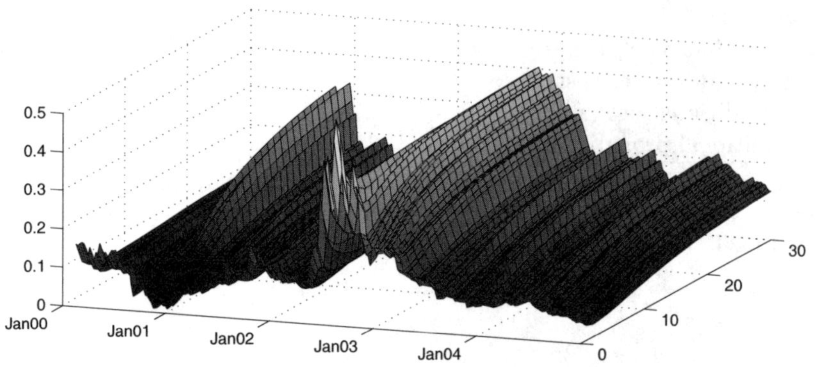

Source: Author's calculations. Term structure implied by weekly estimates of the Gumbel model for maturities of up to 30 years.

Table 5.17. Estimated weekly parameters for the Gumbel model

	2000	2001	2002	2003	2004
α	6.61	5.55	3.37	4.34	5.99
	(0.492)	(0.744)	(1.95)	(0.964)	(0.839)
β	6.28	5.06	4.31	4.12	5.10
	(1.96)	(0.459)	(1.54)	(0.611)	(0.625)
ω	0.292	0.191	0.219	0.383	0.462
	(0.0547)	(0.0594)	(0.0312)	(0.0888)	(0.0523)
N	47	52	52	52	53
mean(n)	5.3	8.4	11.5	14.0	18.4

Source: Author's calculations. The parameter α is the location parameter while β is the scale parameter which determine the curvature of the intensity term structure. ω is the implied RFV fraction. The table shows mean values of the weekly estimations and their standard deviation for Brazil. N is the number of weekly observations per year while n is the average number of bond prices observed each week.

The Gumbel model of the intensity term structure uses the same number of parameters as the Weibull model, although the resulting shapes of the intensity term structure look distinctive. Described by

Table 5.18. In-sample pricing errors for the Gumbel model

	2000	2001	2002	2003	2004	Total
1997 Bra 2027 10.125%	0.570	1.05	0.606	0.878	1.16	0.892
1998 Bra 2008 9.375%	1.04	2.61	2.36	1.72	0.896	1.87
1999 Bra 2009 14.5%	1.16	1.91	2.04	1.13	1.06	1.53
2000 Bra 2007 11.25%	0.464	0.991	2.06	1.83	0.868	1.46
2000 Bra 2030 12.25%	0.640	1.15	1.18	1.87	3.02	1.81
2000 Bra 2020 12.75%	0.996	0.948	1.22	1.91	1.97	1.49
2001 Bra 2006 10.25%	-	1.05	1.29	1.64	2.61	1.76
2001 Bra 2024 8.875%	-	0.788	1.13	2.57	0.914	1.56
2001 Bra 2005 9.625%	-	1.71	1.75	3.21	3.15	2.63
2002 Bra 2008 11.5%	-	-	1.36	1.65	0.99	1.36
2002 Bra 2012 11%	-	-	1.94	1.43	1.23	1.56
2002 Bra 2010 12%	-	-	1.26	1.25	1.66	1.42
2003 Bra 2013 10.25%	-	-	-	1.42	1.12	1.23
2003 Bra 2011 10%	-	-	-	0.942	1.45	1.32
2003 Bra 2007 10%	-	-	-	0.740	1.38	1.17
2003 Bra 2019 8.875%	-	-	-	3.84	2.01	2.43
2003 Bra 2010 9.25%	-	-	-	1.01	1.34	1.29
2004 Bra 2014 10.5%	-	-	-	-	0.431	0.431
2004 Bra 2034 8.25%	-	-	-	-	2.35	2.35
Total	0.888	1.47	1.60	1.82	1.74	1.64

Source: Author's calculations. Means of the RMSE in percent for each of the n bonds per year as well as totals.

the location parameter α and the scale parameter β, the term structures are usually U-shaped and upward sloping. The spike for the short rate is created by the case distinction in (4.41) and has been motivated by empirically observed price discounts, even for maturing issues, during distress. As before, the recovery fraction of face value is simultaneously estimated which repeatedly created some distortions in the early sample period of 2000 where data for only a few bonds are available. This explains the unusually high long rates in Fig. 5.4 around end of 2000. For illustrative reasons, two outliers from November and December 2000 are excluded which slightly alters the residual errors in the following tables.

The location parameter α proxies the distance in years from default and is the mode of the density. The scale parameter β determines the slope of the cumulative density of default because it is proportional to the density's standard deviation. Both parameters are positively correlated. During distress, for instance, α, the proxy for time to default, decreases at the same time as β, a proxy for uncertainty. For Brazil, the correlation coefficient is 0.67 throughout the sample. The course of the estimated parameters nicely describe the 2002 Brazilian crisis, during which α hits a minimum of 0.36 on August 9, 2002 (although the highest EMBIG spread of 2,451 basis points was measured on September 27, 2002). While comparable until 2001, the recovery rate ω appears larger than in the previous models, reaching above the 50% mark for some observations in 2003 and 2004.

Table 5.19. RMSE (%) per maturity bucket and year for the Gumbel model

	2000	2001	2002	2003	2004	Total
0–1 years	-	-	-	-	2.41	2.41
1–5 years	-	1.35	1.91	1.95	1.80	1.83
5–10 years	1.01	1.95	1.79	1.41	1.30	1.53
10+ years	0.763	1.00	1.07	2.04	2.02	1.59

Source: Author's calculations. Means of the RMSE in percent for all N observations of bond prices in each respective category for Brazil.

Table 5.20. RMSE (%) per EMBIG spread and clean market price for the Gumbel model

	0–250 bps	250–500 bps	500-1000 bps	1000+ bps	Total
0%–50%	-	-	-	1.47	1.47
50%–75%	-	-	1.26	2.03	1.87
75%–100%	-	1.62	1.37	1.87	1.48
100%–125%	-	1.52	1.77	2.26	1.71
125% or higher	-	1.70	0.975	-	1.58
Total	-	1.56	1.56	1.94	1.64

Source: Author's calculations. Means of the RMSE in percent for all N observations of bond prices in each respective category. The price category refers to the observed market price without accrued interest. Spreads refer to the JP Morgan EMBIG subindex for Brazil.

The overall in-sample fit, illustrated in Tables 5.18 through 5.20 appears inferior to the Weibull model. Only a few bond issues, such as the problematic 2003 Bra 2019 8.875%, fit better with the Gumbel model. Especially in the beginning of the sample period (where only a few bonds were outstanding), the performance of the Gumbel model is superior to the Weibull model. However, the fit for short maturity issues is definitely worse than in all other models. Looking at short maturity bonds, the model undervalues the respective bond (the 2001 Bra 2005 9.625%) by 2.4% on average. The comparatively large short rates implied by the Gumbel model apparently do not suit this bond very well though it has to be kept in mind that the sample is not very rich in short maturity bonds. Residuals increase only moderately with higher EMBIG spreads, and do not show a pattern with regard to the bonds' price range. This is an advantage compared to the Weibull model. There is, however, some evidence of a relationship between residuals and the price range; the higher (lower) the bond price, the more the model undervalues (overvalues) the bond.

Investment strategies based upon Gumbel price residuals turn out to be inferior to those of the Weibull model and in most aspects to all other models as well. After trading costs, only the long-neutral investment strategy exhibits a slightly higher Sharpe ratio than the passive strategy. The probability measures of predictive ability remain broadly unchanged.

Table 5.21. Performance measures of different trading strategies

Bond	Long-neutral	Long-short	Buy-hold
Compounded return	0.062	0.031	0.13
Simple return	0.065	0.032	0.14
Sharpe ratio	0.99	1.1	0.74
Compounded return after costs	0.054	0.016	0.13
Sharpe ratio after costs	0.82	0.065	0.74
Unconditional probability	0.67	0.52	
Conditional probability	1.1	0.74	

Source: Author's calculations. All returns and Sharpe ratios are based on annualized figures based on trading strategies using price deviations from the Gumbel model for Brazil.

5.1.5 The Lognormal Model With RFV

In this model, two parameters, β and τ, are estimated to determine the shape of the intensity term structure, and ω is—as before—the implied recovery fraction of face value. The resulting term structure, depicted in Fig. 5.5, shows hump shaped curves with comparatively low short rates. As in the other models estimated with endogenous

recovery, the limited number of cross sectional observations causes a
steeply increasing curve with unusually high long rates around the end
of 2000. Analogous to the previous models, three observations within
this period are excluded for illustrative reasons.

Fig. 5.5. Surface of Brazilian intensity term structures

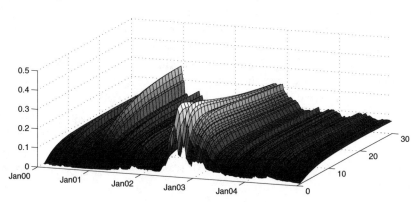

Source: Author's calculations. Term structure implied by weekly estimates of the
lognormal model for maturities of up to 30 years.

The average parameter estimates in Table 5.22 make it more difficult
to infer the country's creditworthiness than, for instance, the Gumbel
model. The implied recovery fractions ω are on average lower than in all
previous models but maintain their relationship to the corresponding
default intensity implied by β and τ.

The overall in-sample fit makes the lognormal model appear superior to the Gumbel and Nelson-Siegel model with exogenous recovery,
but inferior to the two-factor Nelson-Siegel and the Weibull model. In
addition, the model appears to perform better for the last two years
when spreads are exceptionally low.

When looking at single maturity ranges, the very low short rates
implied by the lognormal model offer the best fit to the only bond in
the sample with less than one year to maturity, the 2001 Bra 2005
9.625%. The majority of bond maturities, however, lie in the longer
range where the lognormal model does not perform better than the
two-factor Nelson-Siegel or the Weibull model. Otherwise, errors show
little pattern and give no evidence for systematic biases.

Similar to the two-factor Nelson-Siegel model with endogenous recovery, the lognormal model indicates a better fit for bonds trading at

Table 5.22. Estimated weekly parameters for the lognormal model

	2000	2001	2002	2003	2004
β	0.719	0.806	0.952	0.945	0.980
	(0.145)	(0.106)	(0.115)	(0.0490)	(0.0511)
τ	8.03	7.14	5.04	6.38	9.02
	(0.607)	(0.737)	(2.40)	(1.36)	(1.30)
ω	0.282	0.129	0.134	0.242	0.294
	(0.0563)	(0.0706)	(0.0537)	(0.105)	(0.0636)
N	46	52	52	52	53
mean(n)	5.3	8.4	11.5	14.0	18.4

Source: Author's calculations. The parameter β is the shape parameter and τ the scale parameter which determine the curvature of the intensity term structure. ω is the RFV recovery fraction. The table shows mean values of the weekly estimations and their standard deviation for Brazil. N is the number of weekly observations per year while n is the average number of bond prices observed each week.

high prices (see Table 5.25). This is in contrast to the Gumbel model, and is more comparable to the two-factor Nelson-Siegel model. For different spreads, the model's fitting errors increase with higher default risk, showing a trend comparable to the two-factor Nelson-Siegel.

The model's predictive power in out-of-sample tests show returns in the same range as previous models. The resulting Sharpe ratios are nevertheless lower, especially after accounting for trading costs. The unconditional and conditional probability of correct forecasts are similar to previous models which does not suggest that there is a large predictive element in the residuals.

Table 5.23. In-sample pricing errors for the lognormal model

	2000	2001	2002	2003	2004	Total
1997 Bra 2027 10.125%	0.572	0.948	0.668	1.18	1.25	0.969
1998 Bra 2008 9.375%	1.32	4.30	2.04	1.07	0.609	2.29
1999 Bra 2009 14.5%	1.31	1.50	1.40	0.535	0.508	1.13
2000 Bra 2007 11.25%	0.607	1.53	1.14	1.10	0.704	1.12
2000 Bra 2030 12.25%	0.655	2.08	2.07	0.994	1.02	1.51
2000 Bra 2020 12.75%	1.02	1.08	1.77	0.885	0.75	1.16
2001 Bra 2006 10.25%	-	1.62	1.24	0.778	0.618	1.13
2001 Bra 2024 8.875%	-	0.661	1.05	1.63	0.942	1.15
2001 Bra 2005 9.625%	-	3.82	1.90	0.887	0.400	1.94
2002 Bra 2008 11.5%	-	-	1.20	0.714	0.759	0.896
2002 Bra 2012 11%	-	-	2.03	1.91	0.749	1.65
2002 Bra 2010 12%	-	-	1.31	0.658	0.494	0.834
2003 Bra 2013 10.25%	-	-	-	2.17	1.26	1.63
2003 Bra 2011 10%	-	-	-	0.927	0.769	0.817
2003 Bra 2007 10%	-	-	-	0.796	0.556	0.660
2003 Bra 2019 8.875%	-	-	-	4.59	2.44	2.93
2003 Bra 2010 9.25%	-	-	-	1.06	1.24	1.22
2004 Bra 2014 10.5%	-	-	-	-	0.728	0.728
2004 Bra 2034 8.25%	-	-	-	-	1.21	1.21
Total	1.01	2.23	1.56	1.26	1.01	1.42

Source: Author's calculations. Means of the RMSE in percent for each of the n bonds per year as well as totals.

Table 5.24. RMSE (%) per maturity bucket and year for the lognormal model

	2000	2001	2002	2003	2004	Total
0–1 years	-	-	-	-	0.138	0.138
1–5 years	-	2.70	1.60	0.920	0.636	1.28
5–10 years	1.22	2.78	1.58	1.30	0.894	1.56
10+ years	0.780	1.33	1.50	1.56	1.39	1.38

Source: Author's calculations. Means of the RMSE in percent for all N observations of bond prices in each respective category for Brazil.

Table 5.25. RMSE (%) per EMBIG spread and clean market price for the lognormal model

	0–250 bps	250–500 bps	500-1000 bps	1000+ bps	Total
0%–50%	-	-	-	1.51	1.51
50%–75%	-	-	0.829	1.62	1.47
75%–100%	-	1.50	1.86	1.74	1.83
100%–125%	-	0.707	0.973	0.669	0.905
125% or higher	-	0.472	0.490	-	0.476
Total	-	0.806	1.46	1.65	1.42

Source: Author's calculations. Means of the RMSE in percent for all N observations of bond prices in each respective category. The price category refers to the observed market price without accrued interest. Spreads refer to the JP Morgan EMBIG subindex for Brazil.

Table 5.26. Performance measures of different trading strategies

Bond	Long-neutral	Long-short	Buy-hold
Compounded return	0.072	0.038	0.13
Simple return	0.075	0.039	0.14
Sharpe ratio	1.0	0.95	0.74
Compounded return after costs	0.066	0.026	0.13
Sharpe ratio after costs	0.89	0.46	0.74
Unconditional probability	0.64	0.52	
Conditional probability	1.1	0.89	

Source: Author's calculations. All returns and Sharpe ratios are based on annualized figures based on trading strategies using price deviations from the lognormal model for Brazil.

5.1.6 Discussion

Although the models applied in this section partly result in very distinctive shapes of the term structure, the overall fit compares very well. The classic Nelson-Siegel model has proven its ability to form very distinctive shapes of term structures but performs weakly for distressed bonds. Facilitating the endogenous determination of the RFV fraction presents an appealing approach to significantly improve the models' fit to emerging market sovereign bonds.

Among those models, both the in-sample and out-of-sample performance is broadly homogenous, resulting in closely related recovery rates and similar shapes of the implied pseudo default probabilities. Overall, this result supports the notion that the estimates bear close resemblance to the information contained in bonds and do not remain a synthetic product of the models' restrictions.

The simplified two-factor Nelson-Siegel model presents a variation of the standard Nelson-Siegel model for the simultaneous estimation of recovery fractions. All other models emanate from an unorthodox approach to term structure modeling by using functional forms of probability distributions. Their performance is indeed surprisingly good. Given their parsimonious parametrization, they still yield results even with a cross section as few as five bonds, which indicates a major advantage when analyzing smaller issuers.

The major drawback of the simplified Nelson-Siegel model is the possibility of extreme optimization results under a few circumstances. The reason for this is found in the very restrictive number of term structure shapes the model can form. From the short rate $\beta_0 - \beta_1$, the intensity structure converges to the infinite maturity rate β_0. The rate of convergence is described by the inverse of the parameter γ. No matter which value γ takes, the curve is always concave for $\beta_1 < 0$ and convex for $\beta_1 > 0$. If, for example, under some optimal choice of ω the intensity of short bonds is somewhat lower while the rest of the bond sample shows almost the same intensity, the resulting parameter estimate of γ will become very low while β_1 approaches $-\beta_0$. The resulting short rate approaches zero, its lower boundary. If, for example, the bonds suggest an increasing but convex term structure, the best way for the model to accommodate this are extremely large values for β_0 and γ. This turns the term structure curve into a straight, increasing line. The long-term rate can thereby reach unrealistically high levels. These problems did not emerge as a major obstacle when estimating Brazil's default intensities. However, any model restriction to avoid these caveats resulted in an inferior fit.[3]

Apart from this concern, the two-factor Nelson-Siegel model shows a superior in-sample fit in comparison to the other models. This is unsurprising as it uses one estimation parameter more than the remaining three models. Concerning the out-of-sample fit, the two-factor Nelson-Siegel model does not perform as well. In terms of the risk-return relationship indicated by the Sharpe ratio, the active investment strategy based on a Weibull model exhibits far better performance.

Among the models based on distribution functions, the Weibull model is favored over other distribution models. The lognormal model suffers from its inflexibility of term structure and shows unbalanced error statistics. The Gumbel model creates recovery ratios which are significantly higher than in all other models and indicates an inferior in- and out-of-sample fit compared to the Weibull model. The Weibull model is therefore the model of choice as it shows a sufficiently good and very stable in-sample fit, and performs well as a guideline for an active investment strategy.

5.2 Results From Other Countries

Based on the Weibull model, implied default and recovery parameters are estimated for Argentina, Colombia, Mexico, Turkey, and Venezuela. The results are supplemented by a description of the respective economic circumstances. A summarized analysis of the bond sample and the residuals is included for each country. Space restrictions require condensing the evidence and limiting the number of illustrations.

Data availability dictates the selection of countries. Among those widely researched in related empirical studies, only Russia is not considered here. The Russian Federation has turned to other sources of external financing so that—as opposed to most other countries—the number of outstanding bonds has not increased lately. Due to the very distinctive characteristics of the existing Russian bonds (e.g., by means of put-features), there is no tenable approach to choose a consistent bond sample of sufficient size.

Eastern European countries, such as Hungary or Poland, as well as South Africa use a mix of foreign currency denomination which is less biased towards the US dollar. Of the remaining countries, a large

---

[3] Restricting the long-term rate β_0 to remain below 50%, for instance, doubles the mean RMSE for 2000 and 2001.

number have issued too few global bonds (apart from Brady bonds). This is the case, for instance, for Chile. Other sovereigns may have a sufficient number of bonds, but only small face amounts outstanding. This leads to sticky and uninformative prices, which is the case for Jamaica or Uruguay. Results for Korea did not prove robust since very homogeneous bond prices around par lead to an identification problem as discussed in Sect. 4.4.2. Estimates from the Philippines suffered from a combination of problems (small issue sizes and prices close to par) and are omitted here. Despite these limitations, more countries are expected to be analyzed in the future, as primary markets see more issuance activity and longer time series become available.

5.2.1 Argentina

For the sake of this analysis of current Argentine bonds, the time series of bond prices is truncated as of end of 2001, shortly after the formal declaration of default. The sample contains a small selection of the 152 bond universe on which the government declared default. Many issues, foremost smaller ones, were placed among European investors and hence are denominated in Deutsche mark, Italian lira, or euro. Since prices of smaller issues tend to be sticky and do not convey much useful information, the sample in Table 5.27 contains only US dollar bonds with an issue size of at least \$500 million. This cut-off limit is chosen in accordance with the selection criterium of the Global Emerging Market Bond Index by JP Morgan. Note that the sample also contains bonds issued during the June 2001 mega swap. The exchange reduced the outstanding nominal value of other issues, creating the odd par values in column three. As of the end of 2001, the sample contains 79% of the total outstanding face value from global bonds and represents almost half of all external sovereign debt.

Weekly parameter estimates are illustrated in Fig. 5.6, starting in 1999. During most of 1999, the implied term structure of hazard rates increases sharply since the shape parameter γ almost reaches two. For values between one and two, the shape of the hazard rate term structure is concave. This curvature is more pronounced for lower values of the scale parameter α. When γ declines to slightly above one in autumn 1999, the curve flattens. Since α increases at the same time, the overall term structure is almost horizontal at low hazard rates. In the first half of 2000, hazard rates remain low for all maturities. This comparatively calm period followed the displacement of the decade-long Peronist administration. The initial uncertainty about the significance of the elections (which lead to a democratic transfer of power from the

Table 5.27. Sample of Argentine global bonds

Name	First coupon date	Maturity date	Par ($mln)	Obs	Price range Min	Mean	Max
1993 Arg 2003 US$ 8.375%	20-Jun-1994	20-Dec-2003	2050	149	19.40	55.74	96.85
1996 Arg 2006 US$ 11%	09-Apr-1997	09-Oct-2006	1300	149	19.00	56.84	100.50
1997 Arg 2017 US$ 11.375%	30-Jul-1997	30-Jan-2017	4575	149	18.50	55.00	100.26
1997 Arg 2027 US$ 9.75%	19-Mar-1998	19-Sep-2027	3435	149	17.00	49.76	89.62
1998 Arg 2005 US$ 11%	04-Jun-1999	04-Dec-2005	1000	149	21.00	57.17	100.30
1999 Arg 2019 US$ 12.125%	25-Aug-1999	25-Feb-2019	1433	149	22.35	56.62	105.24
1999 Arg 2009 US$ 11.75%	07-Oct-1999	07-Apr-2009	1500	143	18.50	55.51	102.97
2000 Arg 2020 US$ 12%	01-Aug-2000	01-Feb-2020	1250	94	18.50	46.71	104.60
2000 Arg 2010 US$ 11.375%	15-Sep-2000	15-Mar-2010	1000	94	15.75	46.56	98.60
2000 Arg 2015 US$ 11.75%	15-Dec-2000	15-Jun-2015	2402	81	18.50	44.03	97.25
2000 Arg 2030 US$ 10.25%	21-Jan-2001	21-Jul-2030	1250	81	20.25	39.59	88.28
2001 Arg 2012 US$ 12.375%	21-Aug-2001	21-Feb-2012	1594	43	19.00	35.91	97.63
2001 Arg 2018 US$ 12.25%	19-Dec-2001	19-Jun-2018	7463	28	18.00	29.48	79.25
2001 Arg 2031 US$ 12%	19-Dec-2001	19-Jun-2031	8821	28	18.50	29.23	79.51
2001 Arg 2008 US$ 15.5%	19-Dec-2001	19-Dec-2008	3165	28	18.50	31.72	84.75

Source: Datastream, Bloomberg, Republic of Argentina.

Peronists to the opposition for the first time) might have caused the dip in the recovery fraction ω. Following this, the new administration under Fernando de la Rúa raised hopes for a revitalization of the reform agenda, pulling country spreads down to slightly above 500 basis points. The recovery fraction rebounded to levels around 50% during Argentina's endeavor for sustained IMF assistance which was finally granted in March 2000.

Fig. 5.6. Weekly parameter estimates for Argentina

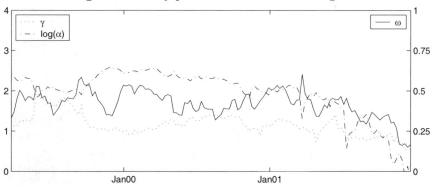

Source: Author's calculations.

While short-term hazard rates remain low during 2000, the overall curve steepens. This results from γ and α moving towards each other, akin to a closing window of opportunity for structural reforms. At the same time, the recovery ratio declines. Fig. 5.7 illustrates the course of credit risk by plotting the implied three year default probability as well as EMBIG spreads. In this graph, the spike in the probability around the end of 2000 coincides with negotiations about an IMF program extension and a rating cut by S&P to BB-.[4] This appears to cause the shape parameter γ to fall, resulting in a flat term structure again while its level shifts to above 10%. The median time to default, calculated by $\alpha(\ln 2)^{1/\gamma}$, falls from approximately four years to two years. When the IMF unveiled the immense size of its aid package in December 2000, markets experienced a clear re-bounce of bond prices during the following weeks.[5] Analogous to EMBIG spreads, the default probability falls by more than ten percentage points from 35%. The recovery ratio ω increases from around 40% to 53%. Between December 2000 and February 2001, the median time to default lengthens from two to three years.

Fig. 5.7. Default probability and EMBIG country spreads

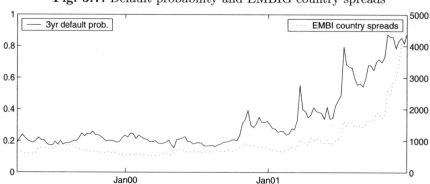

Source: JP Morgan, author's calculations.

In March 2001, the default probability shows a spike, exceeding the 50% mark. Spreads hit 1,000 basis points that week, after contagion from Turkey had already put markets under pressure. Political instability in Argentina resulted in a reshuffle of the cabinet, resulting in the comeback of Peronist Domingo Cavallo as economy minister.

---

[4] The spike in the default probability occured on November 10, the same day as IMF managing director Horst Köhler recommended additional support under the Supplemental Reserve Facility.

[5] The package included IMF financing of 500% of Argentina's quota or $13.7 billion, $6 billion of bi- and multilateral sources, plus additional funds on the order of $20 billion from the private sector.

In May, markets were confronted with plans for a mega bond swap.[6] For a short period, γ reached values far above one, with lower short rates but quickly increasing medium to long-term rates. The transaction was completed in June and succeeded in swapping about \$29 billion of bonds into longer maturities.

The final meltdown began in July when refinancing difficulties in the provinces and rating actions triggered an exodus out of bond markets. The shape parameter γ continues to fall, dropping below one. This indicates an inverted term structure, typical for distressed issuers. The sudden drop of the scale parameter α causes the three year default probability to jump to 80% and the median time to default to decline to below one year. EMBIG spreads reached above 1,500 basis points. The augmentation of the IMF program by a further \$8 billion, which was announced on August 21, causes the default probability to drop to below 60%, followed by an increase of ω back to 37% for a short period. The announcement of a second, quasi coercive debt exchange in November (considered immediately as technical default by S&P) coincides with the final hike of the default probability. The recovery ratio drops below 20%. On November 30th, a bank run started. The final default announcement is correctly reflected in the model with a default probability of one, while ω ends up at 19%.

When the default probability is higher, the recovery of face value determines a larger portion of the total bond value. The finding that ω amounts to a significant number supports the choice of this recovery scheme. This evidence points towards bonds being traded on a "price basis" rather than a "spread basis".

Both the scale parameter α and the recovery fraction ω show a strong and negative correlation to EMBIG country spreads of -0.76 and -0.74, respectively. The RFV fraction's correlation to the cheapest trading bond is 0.79 while the correlation to the mean bond price is slightly lower. This type of relationship is typical for distressed bonds and intensifies with higher spreads. As argued in Singh (2003a) (as well as in Chap. 6), this reflects the perception that the cheapest-to-deliver bond is the best proxy for face recovery, especially under the prevalence of preemptive restructurings without default.[7] Although a preemptive restructuring did not come true in Argentina, the row of debt swaps and continued IMF support might have induced investors to believe in

---

[6] See Chaps. 2 and 3 for more details on the transaction.

[7] It has to be noted that a substantial portion of the bond price for longer maturities is determined by the recovery value. Indeed, the cheapest-to-deliver bond was also the one with the longest maturity most of the time.

a crisis resolution without default. Such hopes were wiped out after the events in November 2001. Thereafter, the correlation between the face recovery fraction and the cheapest bond turns negative (-0.26) indeed.

Table 5.28. Price RMSE (%) per bond and year

	1999	2000	2001	Total
1993 Arg 2003 8.375%	1.91	1.45	2.96	2.22
1996 Arg 2006 11%	1.14	1.36	4.88	3.09
1997 Arg 2017 11.375%	1.03	0.840	3.97	2.49
1997 Arg 2027 9.75%	1.10	0.597	4.39	2.72
1999 Arg 2009 11.75%	1.19	1.57	2.78	2.03
2000 Arg 2030 10.25%	-	1.05	5.49	4.61
2000 Arg 2015 11.75%	-	0.999	6.66	5.37
2000 Arg 2010 11.375%	-	1.40	3.80	2.98
2000 Arg 2020 12%	-	1.19	4.29	3.20
2001 Arg 2018 12.25%	-	-	2.48	2.48
2001 Arg 2012 12.375%	-	-	3.90	3.90
2001 Arg 2031 12%	-	-	5.20	5.20
2001 Arg 2008 15.5%	-	-	3.32	3.32
1998 Arg 2005 11%	1.84	0.627	4.38	2.83
1999 Arg 2019 12.125%	1.77	1.27	7.70	4.77
Total	1.47	1.17	4.71	3.36

Source: Author's calculations.

The in-sample fit is sufficient during 1999 and 2000, but residuals climb higher during 2001. The total RMSE amounts to 3.36%. This proves a better fit than in Berardi and Trova (2003). Errors are homogenous across maturities, although short-term bonds (maturing in less than one year) are not represented in the sample. Shorter bonds appear to trade rich during distress; The total average error of bonds up to a final maturity of five years amounts to -1.6%. When distress intensifies in 2001, however, bonds with very long maturities carry larger but unbiased pricing errors. As can be seen in Table 5.29, the fit decreases significantly with higher spreads and lower bond prices. This is not surprising since the longest bonds trade cheapest as well. The result also reflects that bond trading becomes less liquid by end of 2001.

Merrick (2005) estimates the RFV fraction assuming a different model for the default intensity. His model is closest to the two-factor Nelson-Siegel model used in Sect. 5.1.2, although his decay factor remains constant. This way, the term structure converges smoothly from

Table 5.29. Price RMSE (%) per EMBIG spread and clean market price for Argentina

	0–250 bps	250–500 bps	500-1000 bps	1000+ bps	Total
0%–50%	-	-	-	7.89	7.89
50%–75%	-	-	1.17	4.81	4.64
75%–100%	-	-	1.25	2.38	1.37
100%–125%	-	-	1.33	-	1.33
125% or higher	-	-	-	-	-
Total	-	-	1.26	5.64	3.36

Source: Author's calculations.

an initial short rate to a long-term level. As his model estimates three parameters in almost the same manner, using exactly the same bond sample, results should be closely related. The mean absolute pricing errors, however, appear to be higher by about one-third.[8] The recovery rate in Merrick (2005) is slightly higher at 34%. In turn, his short rate is higher and the term structure shows a steeper decline.

5.2.2 Colombia

The Colombian sample consists of the period after the country's severe recession in 1999.[9] In the 1970s and 1980s, the country enjoyed considerable and comparatively stable growth rates with moderate inflation under a crawling dollar peg. In the 1990s growth accelerated, partly due to capital inflows following a staggered liberalization of the capital account. When these flows reversed in the aftermath of the Asian and Russian crises, the central bank responded with high interest rates, strangling the financial sector and the whole economy. After the devaluation of the peso and a deterioration of the security situation in 1999, GDP contracted by more than 4%. The public sector accumulated considerable external debt as a consequence of recapitalizing the banking system and lost its investment grade rating at the end of 1999. Since September 1999, the central bank has pursued a flexible exchange rate regime with inflation targeting.

The sample allows estimations after February 2000. The 1999 Col 2009 US$ 9.75% was initially issued in April 1999, and in November

---

[8] This compares results for 3-Oct-2001 in Merrick (2005) to this study's result for 5-Oct-2001. Merrick (2005) uses a slightly different optimization function, restricting the average pricing error to be zero.

[9] See IMF Country Report No. 05/154, 14-Apr-2005.

1999 an additional \$500 million were added on. Although Datastream partly reports very distinctive prices for both partial issues, this analysis uses their average price for the sake of a more liquid and consistent time series. This helps to significantly reduce pricing errors. The sample represents only a fraction of all external debt. The global bonds of Table 5.30 make up 20% of all external debt as of end of 2000, and 34% by the end of 2004.

Table 5.30. Sample of Colombian global bonds

Name	First cou-pon date	Maturity date	Par ($mln)	Obs	Price range		
					Min	Mean	Max
1996 Col 2016 US$ 8.7%	15-Aug-1996	15-Feb-2016	200	251	59.50	83.06	104.63
1997 Col 2007 US$ 7.625%	15-Aug-1997	15-Feb-2007	750	251	69.25	94.27	110.08
1997 Col 2027 US$ 8.375%	15-Aug-1997	15-Feb-2027	250	251	59.50	77.71	98.75
1998 Col 2008 US$ 8.625%	01-Oct-1998	01-Apr-2008	500	251	71.00	96.13	113.24
1999 Col 2009 US$ 9.75%	23-Oct-1999	23-Apr-2009	1265[†]	251	74.00	99.17	116.25
2000 Col 2020 US$ 11.75%	25-Aug-2000	25-Feb-2020	1075	251	76.15	103.31	128.93
2001 Col 2011 US$ 9.75%	09-Oct-2001	09-Apr-2011	1000	195	92.25	107.14	117.14
2001 Col 2012 US$ 15%	23-Jul-2002	23-Jan-2012	900	163	80.00	103.20	115.67
2002 Col 2010 US$ 10.5%	09-Jan-2003	09-Jul-2010	507	129	81.13	108.21	119.96
2002 Col 2013 US$ 10.75%	15-Jul-2003	15-Jan-2013	750	108	98.13	111.94	120.10
2003 Col 2033 US$ 10.375%	28-Jul-2003	28-Jan-2033	635	101	91.17	106.00	119.25
2004 Col 2024 US$ 8.125%	21-Nov-2004	21-May-2024	500	50	76.26	89.36	97.44
2004 Col 2014 US$ 8.25%	22-Jun-2005	22-Dec-2014	500	15	99.11	101.96	105.30

Source: Datastream, Bloomberg. † \$765 million issued as of 23-Apr-1999 with an additional \$500 million added on 30-Nov-1999.

The estimated parameters show comparatively low recovery values, especially during 2001 and 2002 (see Fig. 5.8). After the 1999 crisis, an increasing shape parameter γ indicates lower short spreads despite a still increasing, but concave term structure. As of the end of August 2001, γ hits two, resulting in a straight, increasing hazard rate function with a slope of $2/\alpha^2$ or 1.6% per year, respectively. The median time to default remains at six years while ω stays below 20% in the run up to the elections in May 2002. Therefore, default parameters do not evidence a response to failed peace talks with guerilla groups, increased security requirements, and overall lower than expected economic performance. However, a sharp depreciation of the currency lifts country spreads above 1,000 basis points in October 2002 (see Fig. 5.9). The three year default probability shoots up to close to 30%. This change is fueled by overall tensions in Latin America, which were mainly due to the Argentina collapse and concerns about Brazil. The parameters reflect this by a drop in the median time to default to below three years, driven by γ falling to one (which results in a flat hazard rate at around 10%, the inverse of α), and ω close to zero.

Taking office in August 2002, the new administration of Álvaro
Uribe introduced fiscal measures, such as a tax increase and an expen-
diture restraint. While progress appeared slow initially, the situation
improved in 2003–2004. Real GDP grew by 4%, the fiscal deficit came
in at a lower-than-expected 1.3% of GDP, and public debt was reduced
to 53% of GDP by the end of 2004. International reserves covered 123%
of short-term external debt at the end of the sample. This positive development is reflected in a moderate increase in the median time to
default to above five years and significantly higher recovery ratios of
around 40%. The medium-term outlook for the country, however, was
curtailed by depleting oil reserves which made up 25% of exports in
2004. The IMF identified considerable macroeconomic vulnerabilities,
mainly due to the increased level of public debt.[10]

Fig. 5.8. Weekly parameter estimates for Colombia

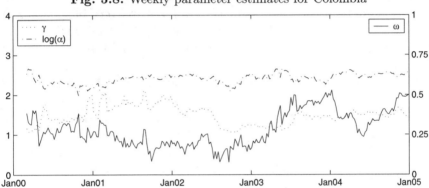

Source: Author's calculations.

The low recovery rates before 2003 hint at the prevalence of the recovery of market value scheme, which is possibly inherent in the pseudo
default probability when ω is close to zero. Since the default probability
is comparatively low for most of the period, however, the recovery value
does not make up a large portion of the bond price. During 2003, external bonds of Colombia were not considered to pose a major debt sustainability problem. Even during the crisis, the maturity structure did
not create pressing rollover needs. Therefore, the parameter results can
be interpreted as evidence that investors assumed market-friendly restructuring terms (if a restructuring-like transaction would be launched

---

[10] See IMF, Colombia Selected Issues, Country Report No. 05/162, 14-Apr-2005, p.
50.

Fig. 5.9. Default probability and EMBIG country spreads

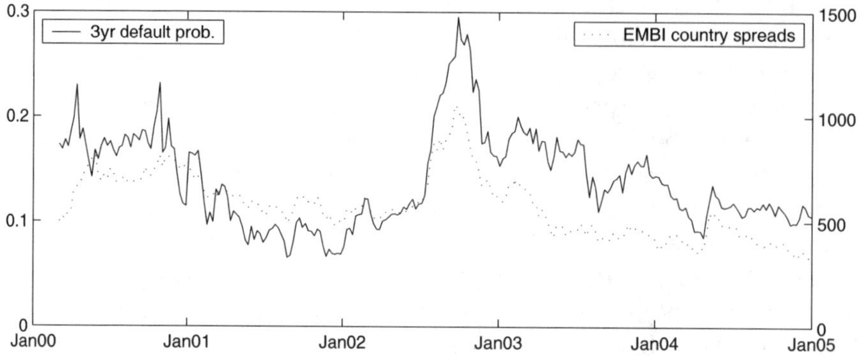

Source: JP Morgan, author's calculations.

at all). The deterioration of debt ratios in 2003, especially with regard to external debt, might have induced a different pattern of recovery in case of a restructuring.[11] This would have required "harder" restructuring terms in order to achieve a considerable debt relief, reflected here by higher values for the RFV fraction ω.

The model fit in Colombia appears inferior to other countries despite the relatively rich data sample and short period of distress. The overall RMSE amounts to 1.95%. There is no pattern across maturities, but pricing errors increase drastically when spreads become high. The high overall pricing error can therefore be traced back to the more turbulent period of 2000–2002. Single bonds can be identified with large pricing errors pointing in opposite directions during certain episodes. For instance, the 1996 Col 2016 8.7% appears overvalued by the model, while the 1999 Col 2009 9.75% is undervalued in 2000. The 2001 Col 2011 9.75% shows consistently higher market prices than the model suggests throughout 2001 and especially during the distress period in 2002. This error is partly compensated for by the 2002 Col 2010 10.5%, which appears overvalued.

5.2.3 Mexico

After the Tequila crisis in 1994/1995, Mexico emerged as one of the best-rated issuers in emerging markets. Since January 2005 the country is rated BBB by S&P. Mexico has also been one of the most active

---

[11] According to JP Morgan, sovereign external debt rose from 28.0% to 30.7% of GDP and total public sector debt increased from 56.7% to 59.4% despite strong GDP growth.

Table 5.31. Price RMSE (%) per EMBIG spread and clean market price for Colombia

	0–250 bps	250–500 bps	500-1000 bps	1000+ bps	Total
0%–50%	-	-	-	-	-
50%–75%	-	0.294	2.05	0.294	2.01
75%–100%	-	1.40	2.39	3.65	2.23
100%–125%	-	1.55	1.96	-	1.68
125% or higher	-	1.36	-	-	1.36
Total	-	1.51	2.23	3.27	1.95

Source: Author's calculations.

issuers in the recent past, widely recognized for its move towards collective action clauses (CACs) in 2003. Today, Mexico is the second largest debtor in JP Morgan's EMBI Global with a weight of 18%.

Mexico's liability management has become very sophisticated. Issuing activities picked up after 1998 with a strong intention to increase maturities, diversify refinancing risks and the investor base, and increase liquidity. Early repayments of loans and the initiation of buybacks of bonds with embedded options were used to take advantage of declining yields. This way, Mexico pre-payed all remaining Brady bonds in July 2003 for a total of $3.5 billion. In April 2004, Mexico initiated an innovative debt exchange with the purpose of improving the efficiency of the yield curve. In a Dutch auction, bondholders of the US dollar issues due 2019, 2022, and 2026 were offered to exchange their bonds into reopened issues of the 2014 or 2033 notes. This design aimed at the elimination of debt trading persistently above the yield curve. Furthermore, the deal increased the share of bonds including CACs. A total amount of $2.3 billion, or 34% of the eligible face value, was exchanged in the transaction.

The sample contains all global bonds issued in US dollars by the United Mexican States. Bonds of the state-owned oil company Petróleos Mexicanos (Pemex), although part of the EMBIG, are not considered here because they are not issued by a sovereign entity. The total sample has a size of $36.6 billion, one of the largest in this study, and represents a 46% share of Mexico's total external sovereign debt or one fifth of all public sector debt.[12]

---

[12] As of end-2004, according to JP Morgan. Mexico's sovereign debt is by a considerable part augmented by debt of other public entities such as state-owned enterprises and a public, extra-budgetary trust fund holding debt emanating from the

Table 5.32. Sample of Mexican global bonds

Name	First cou-pon date	Maturity date	Par ($mln)	Obs	Price range Min	Mean	Max
1996 Mex 2026 US$ 11.5%	15-Nov-1996	15-May-2026	1750	298	105.15	130.16	156.27
1996 Mex 2016 US$ 11.375%	15-Mar-1997	15-Sep-2016	2395	298	101.24	126.41	153.10
1997 Mex 2007 US$ 9.875%	15-Jul-1997	15-Jan-2007	1500	298	97.56	111.57	124.63
1998 Mex 2008 US$ 8.625%	12-Sep-1998	12-Mar-2008	1500	298	90.44	106.88	123.57
1999 Mex 2009 US$ 10.375%	17-Aug-1999	17-Feb-2009	1925	298	96.88	115.40	133.32
1999 Mex 2005 US$ 9.75%	06-Oct-1999	06-Apr-2005	1000	298	97.00	108.03	115.38
2000 Mex 2010 US$ 9.875%	01-Aug-2000	01-Feb-2010	2000	255	97.78	115.80	132.98
2000 Mex 2006 US$ 8.5%	01-Feb-2001	01-Feb-2006	1500	229	98.13	108.24	116.31
2001 Mex 2011 US$ 8.375%	14-Jul-2001	14-Jan-2011	2500	205	96.60	110.40	124.06
2001 Mex 2019 US$ 8.125%	30-Jun-2001	30-Dec-2019	3300	194	89.75	104.98	120.36
2001 Mex 2031 US$ 8.3%	15-Feb-2002	15-Aug-2031	3250	175	89.80	106.11	122.76
2002 Mex 2012 US$ 7.5%	14-Jul-2002	14-Jan-2012	1500	153	97.91	108.46	118.85
2002 Mex 2022 US$ 8%	24-Mar-2003	24-Sep-2022	1750	117	95.38	108.05	118.17
2003 Mex 2013 US$ 6.375%	16-Jul-2003	16-Jan-2013	2000	101	97.59	103.81	111.26
2003 Mex 2015 US$ 6.625%†	03-Sep-2003	03-Mar-2015	2000	94	97.44	104.07	111.83
2003 Mex 2008 US$ 4.625%†	08-Oct-2003	08-Oct-2008	1500	88	97.50	100.99	105.60
2003 Mex 2033 US$ 7.5%†	08-Oct-2003	08-Apr-2033	3057	88	93.69	103.72	113.57
2003 Mex 2014 US$ 5.875%†	15-Jul-2004	15-Jan-2014	1793	62	94.47	99.99	104.45

Source: Datastream, Bloomberg. † Includes collective action clauses.

The comparatively rich sample of bonds, which often trade around par, has created few problems when simultaneously estimating default and recovery parameters. The cross sectional diversity in prices compensates for the lack of default risk when spreads are below 500 basis points. The only problems occurred in 2000, when the sample was relatively small (seven to nine bonds) but spreads contracted to below 400 basis points with prices averaging slightly above par. The resulting ω appears to be comparatively low and very volatile, indicating a potential over-parametrization problem (see Fig. 5.10).

After the peso crisis in 1994–1995, the Mexican economy recovered quickly. As GDP growth was reinstalled in 1997 through 1999, and headline inflation decreased to 4%, bond spreads also reverted from peaks of around 1,000 basis points during the Russian crisis in 1998. At the beginning of the estimation period, the median time to default is about four years at a high recovery rate of more than 50%. This results in an almost flat hazard rate term structure at around 7%. The spikes in the three year default probability are created by γ briefly falling below one (see Fig. 5.11).

In March and April 2000, the recovery rate rapidly falls to less than 20% and remains volatile for the rest of the year with an average of 33%. This move is exceptional but unrelated to changes in the sample (as there are two global bonds issuances in 2000). At the same time,

resolution of the banking system crisis. The figures for the federal government alone are much lower.

Fig. 5.10. Weekly parameter estimates for Mexico

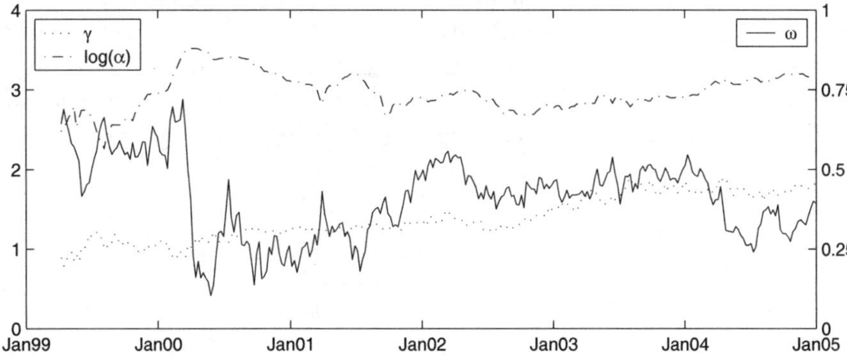

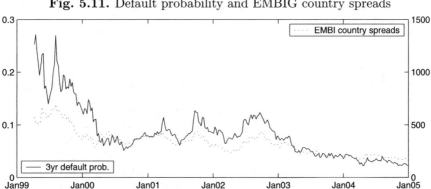

Source: Author's calculations.

Fig. 5.11. Default probability and EMBIG country spreads

Source: JP Morgan, author's calculations.

however, both γ and α surge, resulting in an increase of the median time to default from six to eleven years. The term structure curve, formerly flat or even inverse, is now increasing with lower short rates. This development could mirror surprisingly positive developments in the real sector, as both the March and April growth rates of industrial production of 8.1% and 9.7% year-on-year turned out to be far above consensus expectations.[13]

In the following years, swings of EMBIG country spreads and the default indicators (as implied by the three year default probability) reflected mixed economic circumstances. However, γ and α show clear trends, marking an increasing, concave term structure which is typi-

---

[13] As surveyed by Bloomberg.

cal for high-rated issuers. Both parameters are negatively correlated to country spreads, with γ showing a correlation coefficient of -0.82. Short rates get very low, especially after mid 2003. The recovery rate regains levels above 40% and remains mostly constant in 2002 and 2003. Towards the end of the sample, however, ω declines significantly to between 25% and 35%. As expected for a high-quality debtor, the recovery rate is mostly unrelated to bond prices.

Table 5.33. Price RMSE (%) per maturity bucket and year for Mexico

	1998	1999	2000	2001	2002	2003	2004	Total
0–1 years	-	-	-	-	-	-	0.354	0.354
1–5 years	-	-	1.28	0.820	0.609	0.815	0.392	0.694
5–10 years	-	0.553	0.477	0.603	1.41	0.628	0.519	0.784
10+ years	-	1.48	1.17	0.913	0.683	0.735	0.841	0.883

Source: Author's calculations.

Table 5.34. Relative price error (%) per maturity bucket and year for Mexico

	1998	1999	2000	2001	2002	2003	2004	Mean
0–1 years	-	-	-	-	-	-	-0.325	-0.325
1–5 years	-	-	-1.23	-0.723	-0.461	-0.650	-0.221	-0.504
5–10 years	-	-0.0882	0.109	0.141	-0.0501	0.0750	0.161	0.0679
10+ years	-	0.0919	0.063	0.0408	0.263	0.0735	-0.0724	0.0615

Source: Author's calculations.

The model results in an overall RMSE of 0.8% which is very low (see Table 5.33). Measured by the absolute price errors in relation to the market price, the model's average error is about half of that yielded in Berardi and Trova (2003). Errors increase slightly with maturity as issues below five years remaining maturity appear to be undervalued, while longer issues seem overvalued by the model (see Table 5.34). As before, there is a slight increase of errors with EMBIG country spreads.

Since March 2003, Mexico has included collective action clauses in all of its new global bond issues. Whether the inclusion of CACs results in higher yields (as CACs offer incentives to restructure) or lower yields (as CACs avoid coordination problems in a restructuring) is still a controversial issue in the literature. While theoretical models suggest ambiguous effects, empirical studies are not able to consistently separate the influence. Dixon and Wall (2000), Gugiatti and Richards (2003), or Becker et al. (2003) do not identify a significant influence. Eichengreen and Mody (2000b,c) identify an impact only after controlling for the credit standing. Under their specification, CACs would reduce spreads as Mexico is regarded as a high-quality issuer.[14]

With regard to the CAC puzzle, the Weibull model does not make any effect clearly recognizable. An estimation without the CAC bonds yields similar results. The means of all respective parameters remain broadly unchanged. The correlation of the resulting parameters for the whole sample and the non-CAC subsample ranges from 51% (for γ) to 96% (for α). Smaller parameter deviations, however, can also be a result of the smaller sample.

5.2.4 Turkey

The Turkish sample includes the government's US dollar denominated bonds, leaving aside the country's numerous euro issues. It contains few bonds from the pre-1999 era, but many from refinancing activities after the lira devaluation in 2001 instead. The maturity spectrum is extraordinarily rich with three bonds maturing during the observation period and other maturities reaching up to 30 years. The total nominal volume amounts to $16.3 billion, nearly one fifth of Turkey's sovereign external debt. With this amount of debt, Turkey ranks fourth among the largest index weights in the EMBI Global. The sample excludes a $600 million issue, the 1999 Tur 12% 2008, which was putable at par and traded extraordinarily rich especially during the most severe periods of the crisis.

The estimates gained are depicted in Fig. 5.12. The shape parameter γ indicates an increasing, but concave hazard rate curve for almost all observations.[15] The scale parameter α, proportional to the distribution's standard deviation, varies between six and eight for most of the sample. In autumn 2003, it quickly increases to twelve where it remains for most of 2004. This reflects the country spreads, which halved to around 300 basis points by the end of 2003, supported by rating upgrades. The short-lived reversion to 500 basis points is reflected in a sharp drop of α while the other parameters remain unaffected. The high

---

[14] The sample in Eichengreen and Mody (2000b,c), however, compares New York and London issued bonds between 1991 and 1999, while the circumstances and characteristics of the Mexican CACs issued under New York law are different.

[15] The estimation results of an affine model with recovery of market value in Berardi and Trova (2003) suggest a similarly increasing, but convex term structure of forward spreads.

Table 5.35. Sample of Turkish global bonds

Name	First cou- pon date	Maturity date	Par ($mln)	Obs	Price range Min	Mean	Max
1997 Tur 2002 US$ 10%	23-Nov-1997	23-May-2002	400	132	95.05	100.60	103.13
1997 Tur 2007 US$ 10%	19-Mar-1998	19-Sep-2007	600	269	80.34	100.02	116.47
1998 Tur 2005 US$ 9.875%	23-Aug-1998	23-Feb-2005	400	269	83.50	100.41	108.82
1998 Tur 2003 US$ 8.875%	12-Nov-1998	12-May-2003	300	183	90.59	98.93	104.75
1999 Tur 2009 US$ 12.375%	15-Dec-1999	15-Jun-2009	1250	269	80.80	106.70	129.62
1999 Tur 2004 US$ 11.875%	05-May-2000	05-Nov-2004	500	227	88.90	104.35	110.64
2000 Tur 2030 US$ 11.875%	15-Jul-2000	15-Jan-2030	1500	259	73.89	105.65	145.12
2000 Tur 2010 US$ 11.75%	15-Dec-2000	15-Jun-2010	1500	238	76.88	104.02	129.14
2001 Tur 2006 US$ 11.375%	27-May-2002	27-Nov-2006	1000	162	91.38	109.10	119.30
2002 Tur 2012 US$ 11.5%	23-Jul-2002	23-Jan-2012	1000	154	84.75	109.50	129.77
2002 Tur 2008 US$ 9.875%	19-Sep-2002	19-Mar-2008	1350	146	85.81	104.56	117.32
2002 Tur 2008 US$ 10.5%	13-Jul-2003	13-Jan-2008	1100	112	89.00	109.61	119.23
2003 Tur 2013 US$ 11%	14-Jul-2003	14-Jan-2013	1500	103	88.38	113.57	128.62
2003 Tur 2014 US$ 9.5%	15-Jan-2004	15-Jan-2014	1250	67	96.00	111.26	119.06
2004 Tur 2034 US$ 8%	14-Aug-2004	14-Feb-2034	1500	51	86.00	98.98	106.72
2004 Tur 2011 US$ 9%	30-Dec-2004	30-Jun-2011	750	27	99.47	110.04	114.64
2004 Tur 2015 US$ 7.25%	15-Mar-2005	15-Mar-2015	1500	13	99.13	101.61	104.25

Source: Datastream, Bloomberg.

level of α at the end of the sample causes the term structure of hazard rates to stretch out although the shape parameter γ still indicates a moderately positive slope.

Fig. 5.12. Weekly parameter estimates for Turkey

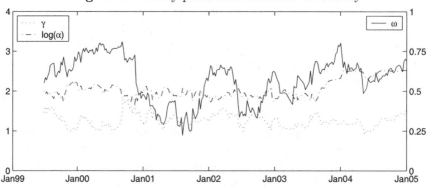

Source: Author's calculations.

The shape parameter γ stays above one for almost the whole sample which translates into increasing, but concave term structure curves. The jump to its maximum in September 2000 occurs shortly before the outbreak of the crisis. When looking at Fig. 5.13, it is striking that the default parameters result in a more or less constant default probability between 1999 and 2003. The median time to default varies between two and three years. As spreads fluctuate between 500 and 1,000 basis points, the change in bond prices can be explained best by movements in the implied recovery of face value fraction. The correlation of recovery rates and spreads is found at -0.83.[16] At spreads around 400 to 500 basis points at the beginning of the observation period, the recovery fraction of face value rises above 60% while falling to below 40% when spreads are above 1,000 basis points. This relationship grows even stronger at high spreads. The same result is shown by the correlation of ω with bond prices which is more pronounced during periods of increased vulnerability. When EMBIG spreads are below their median of 590 basis points, the correlation between the cheapest trading bond and ω is 0.32, while it increases to 0.82 for spreads above the median.

Fig. 5.13. Default probability and EMBIG country spreads

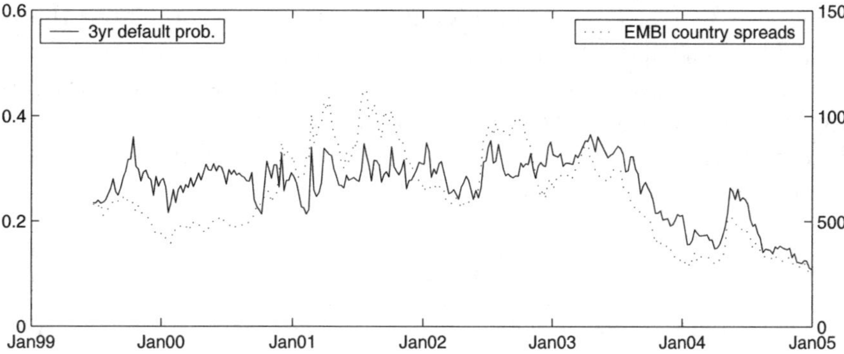

Source: JP Morgan, author's calculations.

As the recovery ratio is the decisive driver of bond prices throughout the period, it is not surprising that ω rises to close to 80%, the highest value in the sample period, by the end of 1999 (and early 2000). After a devastating earthquake in August 1999, Turkey quickly negotiated a $4 billion stand-by credit from the IMF. The recovery fraction ω stabilizes above 40% for a few months after the program was complemented by $7.5 billion from the Supplemental Reserve Facility in December 2000. Its augmentation to a total of $19 billion in May 2001 causes ω to increase from 31% to 44% for some weeks again, before it reverses to its initial value while EMBIG spreads passed the 1,000 basis points mark. However, the situation calmed in the beginning of 2002, when ω

---

[16] This explains also the difference in implied pseudo-probabilities and term structures in comparison to other analyses, such as Rocha and Garcia (2005).

reaches well above 60%. This occurs amid an extension of the IMF's aid package, partly in response to the de facto market closure triggered by the 9-11 terrorist attacks.[17] The same pattern is seen in August 2003 when another extension of the program was granted. The bottom line of this analysis supports the widespread notion that Turkey is mainly a political play due to its strategic importance, keeping its implied default probability below a critical threshold despite weak fundamentals.

The model's fit to the data appears satisfactory. The mean RMSE is comparatively low at 1.2%. The mean percentage price error is significantly lower than in the RMV-based affine model of Berardi and Trova (2003), estimated by Kalman filtering. Due to the large coupons and the absence of spreads far above 1,000 basis points, sovereign bond prices did not completely collapse in Turkey. Pricing errors are homogenous for different price levels although errors are higher, on average, when spreads widen. As Table 5.36 shows, residuals increase with maturity during the distress years 2000 and 2001, but appear unrelated thereafter. The fact that the sample includes issues maturing throughout the observation period does not cause any problems. This characteristic is an advantage of this model in comparison to those models of bond spreads which display distortions since spreads of the maturing bonds show higher volatility.

Table 5.36. Price RMSE (%) per maturity bucket and year for Turkey

	1998	1999	2000	2001	2002	2003	2004	Total
0–1 years	-	-	-	1.06	0.616	0.516	0.838	0.788
1–5 years	-	0.820	1.21	1.21	0.972	1.47	0.750	1.14
5–10 years	-	0.756	1.25	1.98	1.18	0.826	0.662	1.16
10+ years	-	-	0.632	0.149	0.299	1.08	2.74	1.72

Source: Author's calculations.

5.2.5 Venezuela

Venezuela is currently the index' fifth largest issuer with a total weight of 6.0%.[18] In the sample, the 1997 Ven 2027 US$ 9.25% is clearly the dominating benchmark bond with a face value of $4 billion, corresponding to an index weight of 1.76% alone (see Table 5.37). As most of the

[17] Part of the package, about $6 billion of the Supplemental Reserve Facility, would have matured in 2002.
[18] See JP Morgan's Emerging Markets Bond Index Monitor, June 2005.

issuing activity originates from 2003 and 2004, the available time series is comparatively short. For this reason, parameter estimates commence in 2003 and are estimated from an initial sample of only four bonds, each with very distinctive amounts outstanding (see Fig. 5.14).

Table 5.37. Sample of Venezuelan global bonds

Name	First cou- pon date	Maturity date	Par ($mln)	Obs	Price range		
					Min	Mean	Max
1997 Ven 2007 US$ 9.125%	18-Dec-1997	18-Jun-2007	315	57	98.75	104.85	110.00
1997 Ven 2027 US$ 9.25%	15-Mar-1998	15-Sep-2027	4000	57	78.31	92.08	106.26
1998 Ven 2018 US$ 13.625%	15-Feb-1999	15-Aug-2018	500	57	108.66	120.25	136.08
2003 Ven 2010 US$ 5.375%	07-Feb-2004	07-Aug-2010	1500	57	75.13	84.54	94.31
2003 Ven 2018 US$ 7%	01-Jun-2004	01-Dec-2018	1000	57	69.00	79.26	91.67
2004 Ven 2034 US$ 9.375%	13-Jul-2004	13-Jan-2034	1500	51	77.50	92.08	106.26
2004 Ven 2013 US$ 10.75%	19-Sep-2004	19-Sep-2013	1487	40	94.68	107.97	119.65
2004 Ven 2014 US$ 8.5%	08-Apr-2005	08-Oct-2014	1500	12	98.25	103.17	106.09

Source: Datastream, Bloomberg.

The situation in Venezuela is marked by macroeconomic stabilization due to windfall oil revenues, but the medium- to long-term outlook remains uncertain. While country spreads are contracting, the estimated parameters allow a more differentiated analysis of the circumstances.

The country suffered from political instabilities back in 2002, when the opposition launched two strikes to remove president Hugo Chávez from office. Among other effects, this paralyzed the important oil industry in December 2002. In August 2004, Chávez finally defeated a recall which had been lingering in a dispute lasting over one year. This political victory helped to overcome the significant contraction of real GDP in 2002 and 2003. Due to base effects and booming oil revenues, Venezuela's GDP expanded by 17% in real terms during 2004. The following tightening of spreads is reflected in Fig. 5.14 by higher values of both the scale parameter α and the face recovery fraction ω. Median time to default rises from below four years to above five years by the end of 2004. This upswing in bond prices gains momentum after May 2004 but is less pronounced for long-term bonds.

The fact that the longer bonds recover more slowly from their recent lows of around 80 cents on the dollar indicates the vulnerabilities present in Venezuela. This leads to a steepening of the yield curve. The parameter results reflect this by an increase in the shape parameter γ. The resulting term structure of hazard rate remains concave, but indicates a more pronounced upward slope. The highest hazard rate for the longest bond, the 30-year 2004 Ven 2034 US$ 9.375%, is measured on

17-Sep-2004, amounting to 25%. This reflects investor concerns which emanate from the administration's unconventional policy mix. Capital account restrictions—a safeguard for the fixed exchange rate—have indeed helped to double international reserves. These cover almost 100% of all outstanding external government debt. But the government budget has become more vulnerable to future downturns in the oil price. The current windfall oil revenues, making up half of all fiscal revenues, were deployed in a loose fiscal stance which may prove difficult to reverse. Monetary policy appeared similarly abundant as real interest rates were negative despite sustained high inflation. In turn, monetary aggregates and private sector credit were growing at a fast pace. This situation, together with Chávez' rhetoric against capitalism, makes caution well warranted although external debt is not at the center of the concerns.

The overall in-sample fit of the model is high with an average RMSE of 0.71%. Pricing errors show no pattern with regard to remaining maturity. With higher bond prices, however, errors decrease slightly. Since EMBIG country spreads first stay around 600 basis points and fall to 400 basis points by the end of the observation period, there is little variance from which to comment on the errors' relation to country spreads.

Fig. 5.14. Weekly parameter estimates for Venezuela

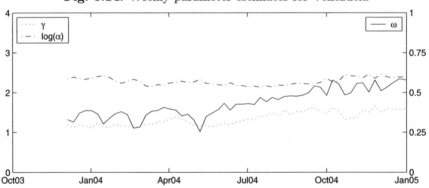

Source: Author's calculations.

5.3 Concluding Remarks

As the case study of Brazil sovereign bonds has shown, all hazard rate models introduced in the previous chapter performed reasonably well in fitting the observed bond prices. However, introducing an endogenous recovery fraction of par facilitates a much better match of bond issues with short remaining life and bond prices during distress. This deviates from the current mantra of modeling bond spreads, which implicitly assumes fractional recovery of market value only. After the economic underpinning provided in the previous two chapters, arguing in favor of a recovery of face value (RFV) framework, the empirical implementation has provided further support for this approach. Even when compromising on the flexibility of the hazard rate model (by choosing a model with fewer parameters), the empirical fit, both in- and out-of-sample, beats the performance of a standard Nelson-Siegel model applied to bond prices.

Among the hazard rate models (using RFV recovery), a simplified two-factor Nelson-Siegel model and three models based on probability distributions perform almost equally well. The goodness-of-fit, the main evaluation criterium used, is very satisfactory given the large range of coupons, maturities, and risk spreads encountered in this case study. For an extended application on five other emerging market sovereigns, the Weibull model is chosen as favorite, given the limited flexibility of the two-factor Nelson-Siegel model (shaping the term structure of hazard rates only by determining the short and the long rate and one convergence factor). While the parsimonious modeling of the term structure represents an innovation (extending the previous work in Andritzky (2005)), the resulting shapes of the term structure are sufficiently flexible and easily applied empirically.

The analysis of global sovereign bonds issued by Argentina, Colombia, Mexico, Turkey, and Venezuela creates a set of implied parameters for the hazard and recovery rates. The estimates appear to describe the economic and political circumstances present in these countries reasonably well. However, out-of-sample tests for Brazil do not lead to the conclusion that the model has strong forecasting abilities. On the other hand, the model's in-sample-fit is very convincing and proves superior to the results of related studies, such as Berardi and Trova (2003) and Merrick (2005).

It remains a task for future research to use the implied parameters of sovereign credit risk gained from this study in economic models of debt sustainability analyses, portfolio selection, or bond trading. Even a first step in this direction, utilizing only the implied three-year default probability and the recovery rate, would present an improvement vis-à-vis the current use of composite indices of bond spreads in such models. Using spreads of credit default swaps (CDS) instead already mitigates this problem since they are traded upon standardized contractual terms and maturities. However, since CDS spreads are also a joint measure of the default and recovery rates, the following chapter extends the current analysis to CDS markets, combining bond and CDS spreads in one framework.

6

Credit Default Swaps

Credit default swaps (CDS) provide the buyer an insurance against certain types of credit events by entitling him to exchange any of the bonds permitted as deliverable against their par value. Unlike bonds, whose risk spreads are assumed to be the product of default risk and loss rate, CDS are par instruments, and their spreads reflect the fractional recovery of the delivered bond's face value. This chapter addresses the implications of this difference and shows the extent to which the recovery assumption matters for determining CDS spreads. A no-arbitrage argument is applied to extract recovery rates from CDS and bond markets, using data from the Brazil crisis in 2002–2003. Results are related to the observation from Chap. 3, that bond prices do not necessarily collapse and recovery values are substantial.

6.1 An Introduction to CDS

In a CDS contract, the protection buyer pays a periodical insurance premium until maturity or a predefined credit event, whichever comes first (see Fig. 6.1). Upon the credit event, the protection buyer receives the difference between the par and the market value of any eligible bond as compensation. Unlike high-yield CDS contracts on corporate bonds in the United States, market convention and standards of the International Swaps and Derivatives Association (Isda) for a typical emerging market CDS contract allow for restructuring as a credit event. Criteria that trigger a credit event include (i) a change in coupon rates, (ii) a change in principal amount, (iii) a postponement of interest or principal payment date, (iv) a change in ranking of priority, and (v) a change in payment of interest or principal to a non-permitted currency.[1]

Fig. 6.1. Cash flow scheme for a credit default swap

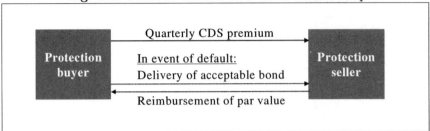

Source: Author's illustration.

In the case of a credit event, the contract is usually settled by physical delivery of the cheapest-to-deliver (CTD) bond against the payment of par value within thirty business days.[2] Since protection buyers do not always have the currently cheapest-to-deliver bond in their portfolio, a distress situation often sets off heavy trading activity where a squeeze in the CTD market may cause such bonds to appreciate. This is the actual recovery value which becomes relevant to the participants in the CDS markets. The recent prevalence of soft restructurings in sovereign markets suggests that this recovery value can become as high as 60%, a level at odds with the analysts' rule-of-thumb of using 25% to value CDS contracts.

This chapter pursues two research questions (see Chap. 2). First, the analysis explores whether the choice of the recovery value has an impact on pricing CDS in a delta-hedge with cash bonds. Duffie (1999a) shows that this is not the case when bond instruments trade close to par, as is common in non-distressed markets. An overly low assumption of the recovery value is offset by an underestimation of the implied default probability, leaving the CDS spreads unaffected. But if prices

---

[1] Permitted currencies are generally the G-7 currencies and also those OECD currencies that carry a rating of AAA on their local currency debt.

[2] According to market participants, physical delivery is the preferred settlement method to avoid price disputes in the typically illiquid cash markets following a credit event. This is problematic when there is a short supply of CTDs, as it recently occurred when the U.S. auto parts maker Delphi went bankrupt in 2005. In this case, an auction to determine the CTD price can be called with protection sellers offering the choice between physical delivery or cash settlement at the auction price.

trade far below par, the offsetting mechanism does not work any longer for two reasons. Firstly, the bond implied default rate increases exponentially with the recovery fraction of par value, as do the CDS spreads. Secondly, since CDS contracts offer insurance for the par value, it is obvious that the insurance premium must be higher than the risk premium implied from bonds which trade below par. The rule-of-thumb that the CDS spread equals the bond yield minus the risk-free rate is not applicable under these circumstances. This obscurity may contribute to the observation that CDS spreads seem to be too high during financial distress.

Regarding the second question, the empirical part extracts information on the recovery value from CDS data which provides a guideline for pricing CDS in future distress situations. In contrast to corporates, the sovereign bond market—with only a limited number of defaults and restructuring events—does not provide enough experience from realized defaults for choosing the correct recovery value in distress. As an alternative, the recovery value can be estimated from bond prices using the method of Chap. 5. As evident from Chap. 3, soft restructurings became the most frequent credit event in sovereign markets. In soft restructurings, bond prices do not collapse when restructurings are announced, but quite the opposite may take place: A comprehensive effort, involving bi- und multilateral lenders, which aims at maintaining debt sustainability, might restore market confidence and lead to lower bond spreads. This results in a much higher recovery rate for CDS contracts than traditionally assumed. Given that the expected recovery value is influencing CDS spreads, the information content of historical CDS spreads can, in combination with the underlying bond prices, be used to reveal the investors' expected recovery value. This chapter shows the result of such an analysis for the Brazil crisis 2002–2003.

This study extends the thriving academic literature on credit default swaps which, among the empirical studies, has largely focused on corporate CDS where the role of recovery is less prominent. Duffie (1999a), Hull and White (2000, 2001), and Schönbucher (2004) present essays on CDS pricing. Empirical studies on the pricing of CDS can be found mainly for the corporate bond market (Cossin et al. (2002), among others), but some working papers recently also research the sovereign bond market (Pan and Singleton (2005), Zhang (2003)). A good overview of the sovereign CDS market and its rapid growth is provided by Packer and Suthiphongchai (2003). The empirical interrelation between CDS and other financial instruments in emerging markets is researched by Chan-Lau and Kim (2004), who get mixed results on whether the CDS market represents the leading source of price discovery. This stands in contrast to studies in the corporate CDS market, where CDS spreads are found to react to new information faster than cash markets (Norden and Weber (2004), Hull et al. (2004), Blanco et al. (2005)). A few other studies have explored the extent of non-default influences on CDS prices. In their analysis of corporate CDS, Longstaff et al. (2004) observe that, besides credit risk, the most important non-default component of CDS spreads is related to liquidity. Packer and Zhu (2005) study the pricing impact of the contractual differences with regard to admitted credit events and deliverable bonds. For the different Isda definitions of credit events, the authors find significant spread differences, both by rating class, as well as by regional and sectoral categories.

6.1.1 CDS Valuation

This section introduces the mathematical foundation of CDS pricing, and explains the reasons why risk spreads might diverge from the cash bond market.

The buyer of protection pays an annual premium of c per one unit of ensured notional value until maturity of the contract, or a predefined credit event, whichever is first. Normally, the premium payments are made quarterly, with the premium payment times denoted by $t_1, ..., t_N$. The first payment accrues and is paid at the end of the first period. The value of these payments, commonly referred to as premium leg, is

$$g(t, c) = \frac{c}{4} \sum_{i=1}^{N} \exp\left(-\int_0^{t_i} r_u + \lambda_u \, du\right),$$ (6.1)

where r_t is the continuous risk-free rate, and λ is the risk-neutral intensity of the credit event. For CDS on assets assumed to bear very high risk, the protection seller may require some upfront compensation. In this case, a lump sum, c_l, in basis points of the underlying face value must be paid at the origination of the contract. The lump sum is added to the protection leg. The upfront payment may be chosen in a way so that the quarterly premiums adopt some odd value, for instance 500 basis points.[3]

In case of a credit event, the buyer of protection receives the full notional amount in exchange for delivery of the underlying bond. Since in

---

[3] When marking CDS contracts to market, this upfront payment reduces the exposure of protection sellers since a smaller fraction of the protection leg is at risk. Therefore, when spreads tighten (widen), the mark-to-market gain (loss) will be smaller.

most cases, sovereigns have a considerable number of deliverable bonds outstanding, the protection buyer has an option to choose, picking the cheapest to deliver bond (on which accrued interest has to be considered). The post-default price of the bond, expressed as fraction of the nominal value, is the relevant recovery value for pricing CDS. Let τ be the default time and τ_d the time of the settlement which occurs within a time span of up to 30 business days after occurrence of the credit event, so that $\tau_d = \tau + d$, with $d = 0, ..., 30$ business days. Resorting to the recovery of face value (RFV) concept, $w(\tau_d)$ is the expected fractional recovery of face value, and $A(\tau_d)$ the value of accrued interest of the delivered bond at time τ_d. Assuming that the protection seller cannot default (i. e. no counterparty risk), define

$$\hat{w}(\tau_d) = w(\tau_d) \exp(-\int_\tau^{\tau_d} r_u \, du) \,,$$

and

$$\hat{A}(\tau_d) = A(\tau_d) \exp(-\int_\tau^{\tau_d} r_u \, du)$$

to signify their corresponding values at the time of default, τ. The present value of the contract for the protection buyer at the time of the credit event τ then becomes

$$1 - \hat{w}(\tau_d) - \hat{A}(\tau_d) - \frac{c^*}{4} \,, \tag{6.2}$$

where $\frac{c^*}{4}$ is the fractional amount of the premium accrued until τ. To calculate the market value of the protection leg at $t_0 = 0$, where the conditional risk-neutral density of a credit event is given by $\lambda(t) \exp(-\int_0^t \lambda(u) \, du)$, the expected value of the protection leg is

$$h(t, c, d) = \int_0^{t_N} (1 - \hat{w}_{u,d} - \hat{A}_{u,d} - \frac{c^*}{4}) \lambda_u \exp(-\int_0^u r_v + \lambda_v \, dv) \, du \,. \tag{6.3}$$

The current fair premium of a CDS contract, $\bar{c}_t$, is chosen so that the value of the premium leg equals the value of the protection leg at the time of origination:

$$g(t, \bar{c}_t) = h(t, \bar{c}_t, d) \,. \tag{6.4}$$

The premium, $\bar{c}_t$, is the spread which is quoted.

6.1.2 The CDS Basis

There are, however, reasons that may lead to a discrepancy between CDS and cash bond spreads. On the one hand, these are conceptual reasons since bond and CDS spreads are calculated differently. The effect of recovery, the main subject of this chapter, is an example. On the other hand, CDS are priced by protection sellers based on their costs to hedge this exposure, resorting to expensive risk capital to cover any residual mismatch. This creates many reasons why in practice CDS might decouple from bond spreads. Such a deviation of CDS from bond spreads is called "basis". The basis is the difference between CDS spreads and the corresponding point on the term structure curve of bond spreads, implying a positive (negative) basis when CDS premiums are above (below) the bond spread curve. Table 6.1 gives an overview of factors which contribute to the basis.

This list of basis effects can easily be expanded, but most of the effects are hard to detect in empirical CDS spreads. The delivery option is usually of little value when one single bond clearly functions as CTD, as is often the case in sovereign bond markets. In the recent distress cases in Argentina, Uruguay, and Brazil, either long maturity bonds or yen denominated issues with characteristically low coupons served as CTD. When particular (small) issues are overwhelmed by demand from protection buyers once distress unfolds, a market squeeze can cause these bonds to appreciate relative to other deliverable issues. This further contributes to the compression of bond prices observed during distress, and bears additional costs to the protection buyer.

Relative liquidity in the cash and protection market is often regarded as an important driver of the basis. In a model exploring the default- and non-default components in bond spreads versus default spreads from CDS, Longstaff et al. (2004) explore the basis for 68 investment-grade rated corporations. Their results suggest that proxies for the liquidity of bonds, such as the bid-ask spread and the principal amount, are the most significant explanatory factors. Packer and Zhu (2005) compare the price impact of different contractual terms with regard to restructurings. Until recently, this has not been relevant for sovereign CDS, since the typical contractual terms include all kinds of restructurings as specified in the introduction. With regard to corporate CDS, the exclusion of restructuring (the "no restructuring (NR)" term) has become more popular today, especially in North America.[4] As expected, Packer and Zhu (2005) find about seven percent higher CDS spreads for

---

[4] Other standard terms, besides full and no restructuring, are (i) "modified restructuring (MR)", first applied in 2001 and for non-sovereigns today by far the most

Table 6.1. Basis effects

	Effect	Description
+	Delivery option	The protection buyer has the choice to deliver any acceptable bond.
	Delivery below par	The protection buyer is insured at par but is only exposed to the trading price of the bond.
	Issuance of new bonds	Pushes up demand for insurance, resulting in a higher price of protection.
	Short selling abilities	If the issuer's credit standing deteriorates, CDS can react more quickly as a short position is more straightforward to arrange.
−	Counterparty risk	Premium compensating for the risk that the protection seller defaults.
	Bond illiquidity	Although the effect can be ambiguous, illiquid paper mostly trades at higher spreads and therefore reduces the respective basis.
	Zero bound	The CDS spread is always positive even if bonds of high-graded issuers trade below the benchmark curve.
	Funding risk	The protection seller does no longer incur funding risk like he would have when replicating the swap by buying the underlying with funds borrowed at the risk-free rate. Repo specialness leads to the opposite.
	Exclusion of credit events	Contractual terms in CDS may restrict the number of trigger event or bonds accepted as delivery.

Source: Author's illustration. Basis effects, contributing to a positive (+) or negative (−) basis, the difference between CDS and bond spreads.

contracts which allow for restructurings as credit event in comparison to NR-contracts. A similar result is gained for their (small) sovereign CDS subsample. No information, however, is given about the absolute impact. For emerging market countries under the auspices of the IMF, the impact might actually be larger.[5] However, to avoid any bias in the

---

popular contract in the U.S., limiting the delivery option to bonds with a maturity of 30 months or less after the maturity of the respective CDS contract; and (ii) "modified-modified restructuring (MMR)", first used in 2003 and often applied in European markets for non-sovereigns, which allows slightly more flexibility in the delivery option than the modified restructuring terms.

[5] According to a model calculation by Merrill Lynch for a CDS spread of 150 basis points, a spread difference of 7% between full and no restructuring contracts implies that only one in five credit events ends in a restructuring with 80% recovery value while all other events would result in a 30% recovery value. See Merrill Lynch, Credit Derivative Handbook 2003, p. 74.

empirical part, this study considers contracts with full restructuring only.

Other basis effects might attract future research in high grade debt markets, but their impact is presumably small and difficult to gauge. The following focuses on exceptional situations of major financial distress, under which these effects were to take a back seat. This will become apparent in the following section which explores the basis that arises when bond prices fall below par while recovery values remain high.

6.1.3 The Role of Recovery

Duffie (1999a) shows that the effect of varying default intensity and recovery fraction is offsetting when pricing CDS in accordance with the underlying bond market. In this case, an underestimation of the recovery fraction will be compensated by an overestimation of the default intensity. When these parameters are used in the CDS spread formula, the bias will cancel out. This mechanism works especially well for short maturities and low par spreads. Nevertheless, it does not work in specific cases which might be more prevalent among sovereign CDS than in the corporate CDS market. Sovereign bonds often show features like low coupons, step-up language, or long maturities which cause bond issues to trade at a wide range of prices. The CTD is mostly trading far below par, especially when spreads are high. For sovereign CDS, there is also a large market for long maturities contracts like ten years, and the contractual terms allow for a wider range of credit events than for corporate CDS.[6] All of these situations mark circumstances in which the offsetting mechanism admittedly fails. This motivates a closer look at the relationship between bond and CDS markets. The following explores the effect of bonds trading apart from par and the related impact of the recovery value. This explains most of the large, positive basis typically observed during distress.

The main driver of the basis is the fact that bond and CDS spreads assume different concepts of recovery, which become incompatible at high spreads. For CDS, the recovery value represents a fraction of the insured nominal of the contract. This risk-neutral recovery fraction, denoted by ω, corresponds to the fractional recovery of the nominal value insured by the CDS contract. This conforms with the RFV concept discussed in Chap. 4. When calculating the basis, the usual proceedure

---

[6] See Packer and Suthiphongchai (2003).

is to assume the same notional value for the CDS and the bonds. However, bond spreads implicitly apply a different measure of recovery. As Chap. 4 exemplifies, bond spreads correspond to the recovery of market value (RMV) concept, where a risk-neutral recovery fraction, ψ, of the pre-default bond value is received upon default.

This discrepancy is best shown when simplifying the CDS and bond spread formulas. Let us assume a constant default intensity, λ, as well as a constant risk-free rate, r, and neglect for a moment the role of accruals et cetera (i.e., $\tau_d = \tau$, $A(\tau_d) = 0$, $c^* = 0$, $c_l = 0$). Imagining continuous premium payments, the formula for the fair CDS premium collapses to

$$\bar{c} = (1 - \omega)\lambda . \tag{6.5}$$

The bond risk premium, as described in Sect. 4.2.4 and according to Duffie and Singleton (1999), is

$$s = (1 - \psi)\lambda . \tag{6.6}$$

Given that both markets are subject to the same default intensity, λ, it is obvious that the two premiums will only be equal (implying a basis of zero) in two cases: (1) the recovery value is zero, so that $\omega = \psi = 0$, or (2) the bond trades close to par. In the latter case, the recovery received as a fraction of face value approximates the recovery received as a fraction of market value, so that $\omega \approx \psi$.[7] The difference between ω and ψ becomes more pronounced when the underlying bond trades farther below par. The corresponding RFV recovery rate, ω, is in this case lower than its RMV counterpart, ψ.

When inferring λ from the bond market under some exogenously given (or assumed) value of ψ (as this is usually done), it is not surprising that the fair CDS spread is higher than the corresponding bond spread when the CTD trades below par.[8] The difference between ψ and ω can have an astonishingly large impact, depending on the respective default risk, maturity, and the assumed recovery rate. This is shown in the following example (see Fig. 6.2). The example assumes risk neutrality, and both a flat risk-free term structure at 3%, as well as a constant default intensity. CDS contracts with one, five, and ten year maturity

---

[7] This is exactly true only if the bond price always stays at par.
[8] The following example illustrates this. If a bond to be insured by a CDS trades at 200 basis points, and ψ is assumed to be 50% homogeneously for all bonds, this results in $\lambda = 400$ bps. If the CTD (regardless of the bond's price the CDS contract is intended to insure) is expected to trade constantly at 0.75 per unit of face value, the relevant recovery rate ω for pricing the CDS becomes $\omega = 50\% \cdot 0.75$, resulting in a CDS spread of 250 basis points.

are priced based on an approximate no-arbitrage relationship with three corresponding bonds, paying a semiannual 9% coupon and maturing in one, five, and ten years. In this simple model, the underlying bonds also serve as the only deliverable bonds. Figure 6.2 illustrates the resulting fair CDS spreads when the underlying bonds trade at spreads of 500, 1,000, and 1,500 basis points.

The graphs show that for zero recovery, the fair CDS spread assumes its intuitively expected value and the basis is zero.[9] For a higher ω, however, the CDS spread becomes very sensitive to changes in the expected recovery fraction. When pricing a five-year CDS contract on the bond maturing in five years, the fair CDS spread remains close to 1,000 basis points for ω remaining in a lower range. This no longer holds when the implied default intensity, λ, grows at increasing pace with higher values for ω. Given a recovery fraction of 0.25, the implied intensity is only 13.7% in the example, but grows to 21.9% and 59.8% for $\omega = 0.5$ and 0.75, respectively. Given high par spreads, the effect of a higher implied default probability arising from a higher ω overrules the effect of decreasing net costs to the protection seller in a default event. This relationship reverses when spreads are as low as 500 basis points and the underlying bond trades above par.[10]

The no-arbitrage relationship used for Fig. 6.2 is flawed by the fact that the respective protection buyer might be overinsured (underinsured) if the underlying bond trades below (above) par. If the protection buyer wants to establish a static hedge of respective bonds at a certain point in time, he might be inclined to buy CDS contracts amounting to a notional which equals his bonds' current market price. This is called partial protection.[11] The above figure, however, assumes that he buys insurance for the full face amount of his bonds regardless of the trading price. For partial protection, the protection buyer would only purchase the proportion of CDS which corresponds to the bond's trading price. This results in a parallel shift of the lines in Fig. 6.2 but does not change their curvature. The overall picture remains the same, even after considering accruals.

---

[9] The slight deviation from the bond spread stems from compounding effects.

[10] As mentioned above, the CDS premium is independent of the recovery ratio when the bond trades exactly at par, which would be the case at a spread of 585 basis points in the example, assuming no accruals.

[11] The corresponding argument for the protection seller requires that, for achieving a neutral position when providing protection for a par value of 100 units of currency, the protection seller would need to short more than 100 units of par value in the underlying bond.

Fig. 6.2. Fair CDS spread as function of ω

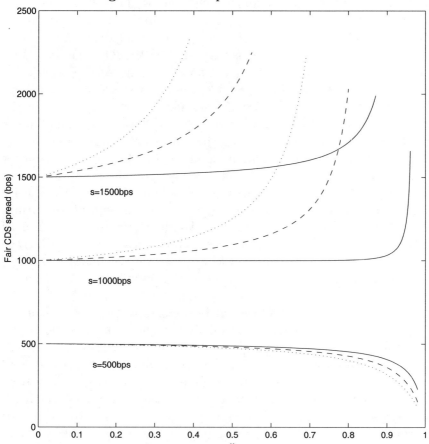

Source: Author's calculation. Fair CDS spreads of one (solid line), five (dashed line), and ten (dotted line) year maturity on a corresponding cash bond with the same maturity and semiannual coupon of 9%. The three sets of lines correspond to the fair CDS spread in dependence of the expected RFV recovery fraction if the bond trades at spreads of 500 basis points (lower set of lines), 1,000 basis points (middle set of lines) and 1,500 basis points (upper set of lines). The risk-free rate is constant at 3%.

How does the picture change when different bonds are acceptable for delivery, not just the one underlying bond as in the above example? This depends on what recovery assumption is applied to the cash bonds. The above relationship does not change when all deliverable bonds share the same recovery value, as suggested by the RFV concept. This concept is most likely to occur under an equalizing restructuring offer, which Chap. 3 has identified as corresponding to hard restructurings. However, if there is a CTD bond available at a lower price than the underlying in the above example (and the relative price difference is maintained after the credit event), the relevant recovery value for the CDS would be the post-default price of the CTD. The RMV concept models this pattern of recovery values by assuming the same fraction, ψ, of the pre-default value to be received. This kind of persistence of price differences throughout a credit event is typical for soft restructurings in which each bond issue is treated differently, mainly for the sake of maturity extension.[12] The question is therefore whether a no-arbitrage argument under the RMV concept leads to the same conclusion, namely that the fair CDS spread is dependent on the assumed recovery fraction ψ.

This question can be addressed by extending the above example with another bond which serves as delivery. The underlying bond and the delivery bond are assumed to have the same expected recovery fraction of market value. Each bond can have a different maturity, and therefore may trade at different par spreads of a non-constant term structure curve. This example creates the same structure of increasing CDS spreads as in the previous setting. Recall that the CDS spread is approximately the product of the default intensity and the loss of face value, $(1 - \omega)$. The default intensity implied by the underlying bond (reflecting the likelihood of the insured credit event) is, for a given bond spread, increasing with the recovery fraction, ψ, as in (6.6). Note that the bond instruments use the RMV version, ψ, while CDS are priced by the fractional recovery of par, ω. The relationship between both parameters, ψ and ω, as gained from the delivery bond, is approximately linear under the assumed circumstances. Their ratio is above (below) one when the CTD trades below (above) par. This creates a picture similar to Fig. 6.2, which shows increasing CDS spreads when the CTD trades below par, and declining spreads when the CTD trades

---

[12] However, as evidence from Chap. 3 shows, even very plain maturity extension deals do not result in a constant ψ for all bonds. In practice, they resemble a mixed recovery framework, the recovery value being the sum of two recovery fractions ω and ψ.

above par. The farther the CTD trades from par, the more the ratio of ψ to ω diverges from one, increasing the flexion of the curves.

As evident from this example, the irrelevance of the chosen recovery fraction is a good working assumption when the deliverable bonds trade near par. In this case, the CDS spread is insensitive to the exogenously chosen recovery value, and the difference between the RMV-based ψ and the RFV-based ω becomes irrelevant. Under such circumstances, other basis effects probably outweigh the recovery effect. This is the reason why, in practice, negative bases are observed for most non-distressed entities, even if the CTD trades slightly below par. In distress, however, the recovery effect cannot be neglected. The pricing of the CTD deserves special attention in such situations. As shown above, this can cause the basis to widen significantly.

6.2 Empirical Evidence From Brazil 2002–2003

The dependence of CDS bases on recovery values offers an opportunity to extract the implied risk neutral recovery value from market data. In the previous chapters, only bond prices were used to estimate the implied RFV fraction. This section will exploit empirical CDS and bond data for the purpose of extracting implied expected default recovery values, leaving aside potential other basis effects.

6.2.1 Preliminary Data Analysis

The empirical analysis relies on different sets of data, based on common data sources (such as Bloomberg and Datastream) as well as quotes directly provided by traders. The analysis of approximate no-arbitrage relationships focuses on episodes of distress, where the recovery effect is most pronounced and dominates other basis effects. Among sovereign issuers, only a few cases are available matching the following selection criteria: (i) a prolonged period of spreads above 1,000 basis points;[13] (ii) a rich data set of CDS quotes; (iii) the existence of cash bonds with roughly comparable tenor to serve as the derivative's underlying bond; and (iv) a set of cheapest-to-deliver bonds trading below par. The Brazil crisis 2002–2003 complies best with these criteria, given the limited availability of liquid CDS quotes from other countries.

---

[13] This threshold is typically seen as a good proxy for distress. See Pescatori and Sy (2004).

Some of Brazil's cash bonds resemble CDS tenors of three and five years, which is useful for the further analysis. The time period of 2002–2003 encompasses the entire cycle of a distress period. Figure 6.3 illustrates CDS spreads and the EMBI country subindex. Credit Trade data indicates that the usual trade size is between $5 and 35 million, with a mean size of $7.5 million. As is typical for distress periods, the term structure of spreads is inverted, with the one year CDS quoting higher than the three and five year contracts. This relation is not reversed until the beginning of 2003, when spreads finally declined. First differences show a significant co-movement of CDS spreads, with correlation coefficients amounting to 96% and 98%, and the first principal component explaining 98.6% of the variation (and 95.6% of the AR(1) residuals).

Fig. 6.3. CDS and EMBI bond spreads during the Brazil crisis 2002–2003

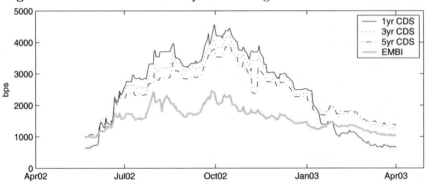

Source: Bloomberg, Datastream. CDS spreads with one, three, and five year tenor and the Brazil EMBI subindex during the Brazil crisis 2002–2003.

Figure 6.3 already gives a clear indication of the development of the basis during this period. Figure 6.4 plots the basis with risk spreads from respective cash bonds of similar tenor. The basis is higher for the three year tenor, which is the result of a less significant inversion of the bond spread curve, both in absolute and in relative terms. Although a number of basis effects listed in Tab. 6.1 might be at work, only a few can be identified by way of offering a conclusive explanation for the basis.

The delivery option does not pretend to bear any value, since the 2001 Bra 2024 US$ 8.875% issue serves as cheapest-to-deliver during most of this period.[14] There is no evidence of extraordinary issuing activity in the cash bond market, and proxies for short selling restrictions prove insignificant. Repo specialness also does not seem to be a main driver of the basis (Longstaff et al. (2004)).

Fig. 6.4. Three and five year CDS basis during the Brazil crisis 2002–2003

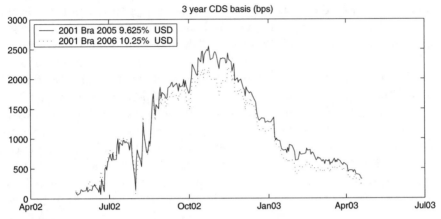

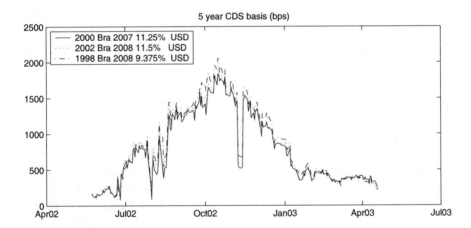

Source: Bloomberg, Datastream, author's calculation. Basis between three and five year CDS and respective cash bonds with similar tenor.

---

[14] On single days, the data suggest lower prices for the 1997 Bra 2006 DEM 11% bond. However, it is believed that these prices are a result of its exceptional small issue size (DEM250 million originally) and would not be maintained if demand increases.

With regard to relative liquidity, Longstaff et al. (2004) show that this effect may widen the basis by around five basis points for liquid bonds, and up to 50 or 75 basis points during distress, according to information retrieved from market traders. For Brazil, indicators of liquidity on the CDS market contribute only marginally to explaining the basis. Generally, Brazil quotes show a mean bid-ask spread of 29.8 basis points, or 15% of the respective mid-quote. This is considerably higher than spreads on Brazil bonds, amounting to around 20 basis points.[15] However, during distress, bid-ask spreads are relatively lower and make up only around 3% of the CDS mid-quote.

The co-movement of the basis to common proxies of risk aversion is less obvious than in the sample countries used by Pan and Singleton (2005). In a regression of levels, the VIX volatility index explains between 10% and 20% of the variance. Other measures, such as the EMBI Global, or the term spread of U.S. Treasuries, do not provide meaningful determinants of the CDS basis.[16] The most significant influence can be traced back to the fact that the underlying bonds traded below par. A regression of the basis on the price difference to par explains more than 40% of the variance. If it is assumed that protection buyers ask only for partial protection (and therefore buy CDS to insure only the current market price of the underlying bonds), the basis is positive solely in autumn 2002, reaching to above 500 basis points. This preliminary analysis indicates that the price effect of these underlying bonds, trading below par, is the main cause for the discrepancy of spreads between the CDS and the cash bond market.

6.2.2 No Arbitrage With Two Instruments

The analysis of basis effects in the previous section has offered several possible explanations, and suggests that the spread difference between CDS and bonds might not be a good measure by which to assess the relative value of a position. The following section calculates the implied default intensity and recovery rate, using a simple no-arbitrage framework. For this reason, bonds of roughly the same tenor as an available CDS contract are selected from the data. Assuming that the CDS is priced in the absence of arbitrage, the observed market values can be used to calibrate the CDS and the bond valuation formulas (see (6.4), and respectively, (4.9) in Chap. 4). Leaving other basis effects aside,

---

[15] As measured by the British Bankers Association according to the EMBI+ subindex.

[16] In fact, their correlation to the basis is negative.

the CDS spread is a function of the default intensity and the recovery fraction,

$$c_t = f(\lambda_t, \omega_t) \, , \qquad (6.7)$$

with the default intensity and recovery rate assumed to be constant throughout the life of the contract. The dirty bond price, in turn, is, under this no-arbitrage argument, a function of the same parameters,

$$P_t = w_{RFV}(\lambda_t, \omega_t) \, . \qquad (6.8)$$

When using a combination of one CDS spread for c in (6.7) and one corresponding bond price for P in (6.8), both variables, λ and ω, can be determined. Figures 6.5 and 6.6 illustrate this implied recovery fraction of face, ω, the main variable of interest, for different combinations of bonds and CDS.

Fig. 6.5. Daily recovery rates implied from three year CDS and cash bonds

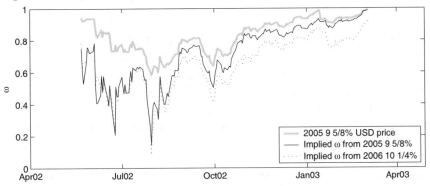

Data source: Author's calculation. Implied recovery rate ω from an approximate no-arbitrage relationship calibrated to three year CDS and bond market data in Brazil 2002–2003. The implied parameters, the default intensity λ and the recovery rate ω, are assumed to be constant for the remaining life of the instruments. The grey line indicates the dirty price of one of the underlyings.

It has to be borne in mind, though, that in this analysis the two implied parameters subsume all statistical noise, residual basis effects, as well as potential distortions emanating from a less than perfect match of maturities. Furthermore, the implied parameters represent risk neutral measures, so that the implied recovery rate does not necessarily compare with realized recovery values from Chap. 3. However, it is striking indeed, that the implied recovery rate during distress is much higher than the 25% standard assumption. The course of the implied

Fig. 6.6. Daily recovery rates implied from five year CDS and cash bonds

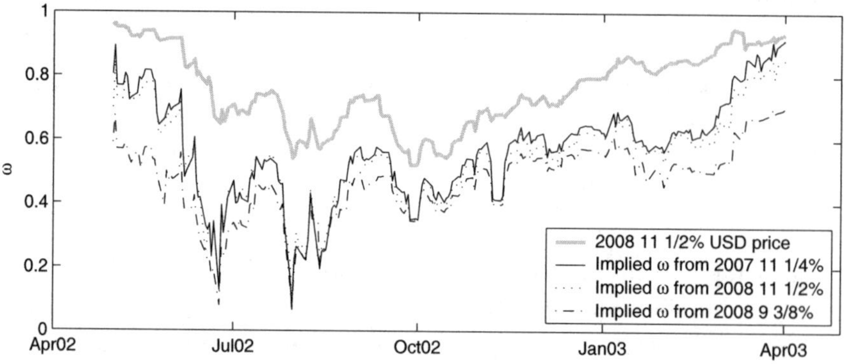

Data source: Author's calculation. Implied recovery rate ω from an approximate no-arbitrage relationship calibrated to five year CDS and bond market data in Brazil 2002–2003. The implied parameters, the default intensity λ and the recovery rate ω, are assumed to be constant for the remaining life of the instruments. The grey line indicates the dirty price of one of the underlyings.

recovery rates shows a strong correlation with the overall bond price level. This is a reasonable result as the underlying cash bond price provides an upper bound for the recovery value. When translating the implied recovery fraction of par into the respective recovery fraction of the underlying bond's market value, the resulting ψ is found to be between 0.74 and 0.80 for the five year contract, and close to 0.9 for the three year contract.

At a first glance, there are two explanations for the apparently very high recovery ratios. First, it could be argued that high recovery values are typical since a soft resolution of debt problems is implicitly guaranteed by the international financial institutions and the G-8 governments. This might be true for Brazil, which at that point received considerable support (see Sect. 2.2.4). The cases of Russia and Argentina, however, have shown that such a conjecture might not be accurate. Second, the result might point towards a discrepancy between market CDS spreads and the theoretical spreads warranted by the model, possibly caused by other basis effects at work. When recalling that the recovery effect is less pronounced at shorter maturities, this conclusion appears even more plausible for Brazil where the recovery rate implied from three year contracts is higher than for five year contracts. It is reasonable to imagine that CDS spreads during very severe distress are higher than the model suggests since certain basis effects (such as a shortfall in the supply of protection and costs of short selling) become more significant. It is also possible that higher spreads reduce the number of protection sellers in the market, creating a quasi-monopoly for very few, well diversified actors.

However, if this implied measure represents the true risk-neutral RFV fraction for all bonds, we would expect the recovery fraction to be about the same for all maturities. Such an interpretation neglects the fact that the implied recovery rates become higher than the CTD's price which, of course, has not yet entered the arbitrage relationship. This gives rise to the more flexible model of mixed recovery where a bond's total recovery value is comprised of the sum of RMV and RFV fractions. This is the most sensible way to draft a recovery scheme which fits all bonds, both the underlyings and the cheapest-to-deliver bond. Under this type of model, the recovery fraction of face value would be lower than the implied recovery value in the above estimations. The total recovery value of the CTD (which is lower than the recovery rate in the above figures) would cause the implied default intensity for the CDS and the underlying bond to assume lower values. This corrects a potential bias emanating from the above assumption of the underlying bond serving as delivery, and sheds more light on the crucial role of the CTD.

6.2.3 No Arbitrage With Three Instruments

This no-arbitrage relationship is made up by three instruments, the underlying bond, the cheapest-to-deliver bond, and the CDS contract. Assuming that our data presents fair prices for each instrument, and other basis effects are negligible, the formulas for CDS and bond prices suggest the distinction of the following unobservable variables: the RMV fraction, ψ, the RFV fraction, ω, and the intensity of a credit event, λ. While term structure models can be used to accommodate different shapes of forward intensities, the following uses a constant hazard rate function (see Sect. 4.3.2), which, together with the flexible model of mixed recovery, presents a sensible simplification suitable to this static estimation. The use of a constant hazard rate is necessary to infer, along with the default intensity, both implied recovery parameters, ω and ψ, from the daily prices of the three instruments under consideration.

The resulting recovery value of the CDS, comprised by the RFV fraction, ω, and the ψ-fraction of the CTD price, are shown for the three and five year contracts in Figs. 6.7 and 6.8.[17] Although it is well known

---

[17] These figures are created for illustrative reasons (instead of plotting solely ω and ψ), and incorporate additional assumptions. The graphed recovery value is

that risk term structures are significantly inverted during distress, using constant default intensities is a reasonable working assumption for two reasons. First, the introduction of face value recovery (in contrast to market value recovery implicitly assumed when looking at spread term structures) causes the implied default intensity to level out as can be seen in a bootstrapping analysis (as illustrated in Fig. 4.7 in Chap. 4). Second, distressed bond prices show little sensitivity to the long-term (survival contingent) default intensity as the likelihood of survival in the distant future is very low. The price of the Brazil CTD, which has a much longer duration than the three and five year tenor contracts regarded here, is therefore only marginally affected by an incorrect specification of the long-term default intensity.

Fig. 6.7. Daily RFV and RMV recovery rates implied from three year CDS, underlying, and CTD

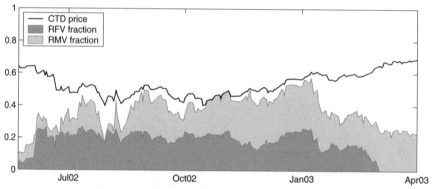

Data source: Author's calculation. Dirty CTD price and implied recovery value of a three year CDS contract, split into RFV and RMV fractions. The Brazil 2006 US$ 10.25% bond serves as underlying.

The Figs. 6.7 and 6.8 show two main results. First, the implied recovery value is found significantly higher than the 25% fraction of

---

discounted with the risk-free rate from the 75% quantile default time, i.e. the point in the future at which the cumulative default probability hits 75%. The recovery fraction of market value, ψ, is multiplied with the current clean price of the CTD which does not take into account fluctuations in the expected pre-default price. Results are illustrated only for one underlying. For the three year tenor, Brazil 2006 US$ 10.25% is the most liquid bond and has the best maturity match. Results for the Brazil 2005 US$ 9.625% bond look comparable, despite an earlier convergence of the RFV fraction to zero at the end of the sample. For the five year tenor, results for the other bonds look similar.

Fig. 6.8. Daily RFV and RMV recovery rates implied from five year CDS, underlying, and CTD

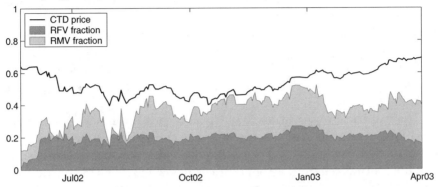

Data source: Author's calculation. Dirty CTD price and implied recovery value of a five year CDS contract, split into RFV and RMV fractions. The Brazil 2008 US$ 11.5% bond serves as underlying.

face most of the time. Second, the graph unveils a very high correlation between the implied recovery value and the dirty price of the relevant CTD during the heights of the crisis. During the period from July 2002 to January 2003, the correlation amounts to 85% for the three year, and 83% for the five year contract.

The level and course of the implied recovery rates are fairly similar for both CDS maturities, except for the ω-recovery fraction which turns to zero in March 2003 for the shorter contract. For the five year contract, the recovery of face value portion, ω, remains stable around 20% of face value while almost all variation originates from fluctuations in the ψ-recovery fraction. During the period of autumn 2002, when spreads hit their peak, the implied total recovery value almost equals the CTD price. This implied recovery value can be interpreted as the investors' expectations with regard to the value recovered from a possible restructuring. The constellation suggests that CDS markets expected an imminent, but soft restructuring. Given the nature of the Brazil crisis and the multilateral support the country enjoyed, the soft restructuring scenario presents an appealing interpretation of the result.

6.3 Concluding Remarks

The conjecture that recovery is not a relevant determinant of CDS spreads is a reasonable approximation during non-distress periods. The theoretical section of this chapter, however, indicates that when bonds trade considerably below par, for instance during distress, the role of recovery cannot be put aside. Awareness of this phenomenon is important for understanding CDS spreads and a well functioning market for protection.

The empirical observation that CDS spreads are much higher than bond spreads during distress exceeds the scope of traditional basis effects (such as the delivery option), and is based on two mechanisms. The first is the fact that, at bond prices below par, a CDS contract insuring the par value offers overinsurance, resulting in a positive basis. Additionally, the relevant recovery rate for CDS contracts is the post-default value of the bond chosen for delivery by the protection buyer. The protection buyer would, preferably, chose the cheapest-to-deliver bond. Independent of the underlying to be protected, the cheaper the expected post-default price of the CTD in relation to the underlying bond, the higher the wedge between the CDS and the bond spread.

The second mechanism refers more directly to the impact of the chosen recovery fraction to price CDS. This is best illustrated in a model calculation in which a CDS contract price is based on an underlying bond with the same tenor, being the only accepted delivery. Using simplified formulas, the bond spread equals the product of the hazard rate, λ, and the loss rate using the RMV definition, $(1 - \psi)$. In contrast, CDS are par instruments so that the CDS spread is the product of λ and the loss rate using the RFV definition, $(1 - \omega)$. At high spreads during distress, the underlying bond usually trades below par. Thus, any exogenously assumed recovery fraction of par translates into a higher recovery fraction of market value, and $\psi > \omega$. Assuming consistent pricing of CDS and bonds under risk neutrality, this causes the fair CDS spread to increase exponentially with higher values for ω, given the bond price is below par. This non-linear relationship becomes more pronounced for higher spreads and longer CDS maturities.

Especially during crises, the above effect can be used to infer the implied recovery rate from market data of CDS and bond spreads, assuming that both are driven by the same parameters. Using data from the Brazil crisis in 2002–2003, this helps to explain why the CDS basis reached levels as high as 2,500 basis points. Calibrating bond and CDS formulas for different combinations of three and five year tenor instruments suggests that the implied recovery rate is closely related to the underlying bond price and remained above 40% of par during most of the crisis. Using a combination of three instruments (the CDS, the underlying bond, and the CTD), it is possible to distinguish between the RMV and RFV fractions, following the mixed recovery model of Sect. 4.4.2. Results for the Brazil crisis show that CDS and bond data implied an almost constant recovery fraction of par (about 20%), while the RMV-fraction adds another portion of up to 25% to the total implied recovery value. The implied recovery value is strongly correlated with the CTD during the heights of the crisis, supporting the idea that the CTD is a useful proxy for pricing CDS (Singh (2003a)). Furthermore, the very high recovery value (in comparison to the CTD) suggests that markets expected an impeding credit event, such as a restructuring at soft terms, in fall 2002.

7

Conclusion

The global bond market for emerging sovereigns has weathered a significant number of adverse events in its recent past. Nevertheless, it is currently prospering due to the benign constellation of the global economy and the improvement of domestic policies. Despite setting a precedent, the historic dimension of the Argentine default and restructuring did not sound the death knell for the sovereign bond market, as some commentators had foretold. The "global saving glut", together with a positive change in attitude towards fiscal prudence in emerging market sovereigns, still motivates large capital flows into the global sovereign bond market. Historically low levels of country spreads not only reflect excess liquidity, however, but also improved fundamentals. Besides reaching higher levels of economic maturation, many governments increased public savings, introduced more flexibel exchange rates, and took advantage of the non-inflationary expansion in global liquidity which pushed down the equilibrium real interest rate.

This monograph has sought to present a holistic picture of the international sovereign bond market from the point of view of an enlightened investor. The historic perspective on sovereign borrowing and lending has shown that, even during the sovereign bond market's early bloom more than one century ago, sovereign lending could prosper in absence of international bankruptcy regulation. The former era also enjoyed less volatile and generally lower spreads despite lower transparency and skewed fundamentals. This look at history sheds a different light on the exceptionally low risk spreads enjoyed by sovereign debtors in the years after 2004.

Yet past cycles of sovereign lending and default have taught that debt crises will recur at some point. Since the fall of Russia, investors know that the "too large to fail" paradigm is invalid. The relatively benign restructuring terms in the test case of Pakistan avoided scaring off investors and thus paved the way for other workouts. Lacking the reputation of Pakistan, Ecuador subsequently moved forward with the first comprehensive bond restructuring on less favorable terms. Given significant external bonded debt, sovereign bond restructurings have become an integral part of IMF programs since. The workouts in Ukraine, Uruguay, and the Caribbean countries make a case for expecting more sovereign bond restructurings in future crises.

The emergence of new crises will, however, be countered by more and more sophisticated policies for crisis containment. For formulating an appropriate policy response to debt crises, it has proven helpful to distinguish situations of illiquidity from insolvency. Sovereign governments in need of financing will always try to tap the source with the lowest marginal cost of borrowing. If all sources are exploited, all stakeholders have to assume responsibility. This results in an early call for private sector involvement, which may take the form of a voluntary debt swap to prolong maturities, a soft or a hard restructuring. While a soft restructuring is a comprehensive take-it-or-leave-it offer to prolong maturities of sovereign bonds for immediate cash-flow relief, a continuum exists towards hard restructurings which additionally ask for debt relief. This continuum mirrors the flexibility needed to counter sovereign debt crises, since fighting crisis symptoms may turn a liquidity problem into a more profound solvency problem.

Sovereign restructurings show, despite the individual circumstances and the prevalence of ad hoc solutions, more commonalties than widely recognized. The large positive return investors could achieve during restructurings indicates the existence of a significant risk premium for these transactions. Such a risk premium appears unjustified given the positive track record of most sovereign restructurings: all transactions were successfully closed with large participation rates, and sovereign bond prices quickly rebounded in most cases. Sovereign restructurings have not justified the fears investors have about them. This lesson grows more important as the international financial institutions call for immediate involvement of private investors during a crisis. Preemptive restructurings, as seen in Uruguay, reduce the criticism of bailing out private investors and ensure timely action for crisis containment. This trend suggests that investors should reckon with more frequent credit events, albeit ones that apply soft terms and result in small losses, if any.

While the "rush to the exits" in the run-up of a restructuring only locks up the losses, which are worst in early stages of the process, the "rush to the courthouse" of holdout investors has little appeal as well. Legal action, at least from smaller bondholders, has not proven to exert significant influence on the debtor and does not meet with political approval. The recent attempts to attach payments to other bondholders, who have accepted the restructuring offer, are seen as an obstacle for successful crisis resolutions. Furthermore, the spreading use of collective action clauses will eliminate the holdout strategy when minority bondholders can be bound into the restructuring deal. Holdouts will soon become history.

The preliminary analysis of recent restructuring deals inspires the design of a bond valuation model to better reflect the peculiarities of sovereign bonds. Tracing the distinction of hard and soft restructurings, the model combines the two major recovery concepts of recovery of face value (RFV) and recovery of market value (RMV). The value of a cross section of sovereign bonds with similar seniority is therefore comprised of a risk-neutral default probability (which can be modeled as a stochastic process) and a mixture of both the RFV and RMV recovery values.

Not only is this model of recovery very well suited to sovereign credit events, but it also applies a new approach for modeling the default probability. To determine the shape of the default intensity term structure, the functional forms of continuous probability distributions are used. This approach presents an extension of Andritzky (2005) and resembles failure rate models from technical engineering. In comparison to traditional parsimonious models, like the Nelson-Siegel (1987) model, such failure rate models fit empirical bond prices much better, even for maturing issues and during financial distress. Given that the bonds trade significantly apart from par, the parsimonious parametrization allows the estimation of implied default variables for countries with as few as five outstanding bonds. In combination with the simultaneous estimation of the recovery fraction of face value, the model is sufficiently flexible to fit a heterogenous cross section of bond prices while avoiding the over-parametrization problem encountered by other models.

In the empirical application, the performance of a Nelson-Siegel model is compared with bond models using different distribution functions for the default intensity and RFV recovery. Based on data from Brazilian sovereign bonds, the in- and out-of-sample fit of the latter type of models is much better. Therefore, the most flexible Weibull model (using the function of the Weibull distribution) is chosen to estimate implied credit risk parameters for half a dozen emerging market countries. Using a cross section of pari passu ranking bonds, the estimates yield a term structure curve of risk-neutral default probabilities and the RFV recovery fraction at any point in time. The resulting weekly parameters neatly describe the economic events during the observation period.

The above provides a useful contribution to the literature in two important ways. First, the model performance can be compared under different parsimonious hazard rate models applied to emerging market sovereigns. The result extends the existing literature by providing a comparative study on the empirical performance of bond pricing models. Second, the resulting parameter estimates yield a more differentiated picture of sovereign credit risk than the ones currently used in the academic literature. Future research can rely on these parameters for cross-country comparisons of sovereign creditworthiness, the calibration of fundamental models of debt sustainability, and the drafting of appropriate crisis resolutions.

The distinction of recovery value modeled as a fraction of face value versus pre-default market value is also relevant for recovery-contingent derivatives, such as credit default swaps. CDS are popular credit derivatives which, due to their standardization, can be used to measure and manage credit risk exposure. The analysis serves as a reminder that CDS and bond spreads are fundamentally different because their pricing formulas use different recovery rate definitions. The difference is most pronounced during distress and highlights the relevance of recovery assumptions. Using data from the Brazil crisis 2002–2003, it can be shown that this fundamental difference explains the large positive basis between CDS and bond spreads. This additional information can be exploited by means of the no-arbitrage relationship between bonds and CDS, yielding insight into the market expectations about the terms of an impending restructuring.

With regard to policy implications, this study adds to the literature that tries to shape the direction in which the international financial architecture is heading. The literature has, to date, substantially contributed to the determination of optimal debt levels and the overcoming of incentive problems. As a result we understand much better when multilateral assistance is appropriate and how emergency lending should be designed without creating misleading incentives. Concerns that IMF stand-by assistance creates moral hazard are unwarranted. Many commentators have pointed out that as long as crisis lending is repaid in full, there is little substance to the argument of bail-outs creating creditor moral hazard. By involving private investors through sovereign bond restructurings, all parties will contribute to the crisis prevention and resolution effort. With regard to debtor moral hazard, most crises go hand in hand with political instability. This connection does not support the notion that a sovereign default serves the current governing elite and is in their own interest. Although the IMF is often compared to a fire brigade, by serving as "scapegoat" instead the institution actually can facilitate indispensable policy change.

Lack of political consensus, however, did prevent the IMF from assuming the role of an international sovereign bankruptcy court. While the discussion, recently fueled by the draft for a "Sovereign Debt Restructuring Mechanism", has increased the awareness of this topic, the establishment of such a centralized mechanism is off the table. The new market standard of bond contracts includes collective action clauses which overcome the collective action problem, facilitate the involvement of bondholders in the negotiations, and reduce the large costs of sovereign debt crises. With majority voting in place, the current policy mantra of preemptive soft restructurings appears as a sensible response to the criticism brought against the international financial architecture in the 1990s. The IMF, however, should go even one step further by tolerating sovereign default where appropriate in order to signal the need for debt relief. This extra step would present the recognition of accepted custom, given that all past sovereign restructurings (except for Uruguay) took place under default, a delay in debt service, or at least a credible default threat.

While crisis prevention and cure hint at the risks inherent in sovereign borrowing, public debt nonetheless is an effective device for disciplining domestic policy and therefore creates credible signals. For instance, by using foreign currency bonds as a lending device, the country imposes on itself constraints with regard to monetary and exchange rate policies. Where securitized debt dominates emerging market finance, markets give immediate feedback on policy actions. This feedback mechanism fosters more sensible policies, as an unsustainable policy mix is quickly revealed and sanctioned by the international markets. Therefore, emerging countries should be encouraged to take advantage of the liquidity in the international financial markets in manifold ways. Besides promoting the issuance activity of new sovereign and sub-sovereign entities, the debtors should be allowed to issue in their domestic currencies, for instance, by a concerted action to overcome the original sin problem. Following the spirit of the Brady plan, such an initiative could help create an international market for domestic currency debt. Investors should be prepared for, and take advantage of, these kinds of innovative deals in the sovereign bond market.

Similarly, the credit derivative market requires more attention by regulators. With the prevalence of soft credit events, CDS spreads become prohibitively high, diluting their capacity as standardized insurance against, and measure of, credit risk. This situation calls for contracts that limit the choice of the deliverable bonds, or reduce the scope of insured credit events (although the distinction of default and restructuring in the sovereign context is ambiguous). The goal should also be to avoid a squeeze in the market for deliverable bonds, as might be the case if the cheapest-to-deliver bond is overwhelmed by demand from protection buyers. The small number of protection sellers who remain in distressed markets might distort the price of protection and lead to the accumulation of delivered bonds of some particular series in the hands of the protection seller, thus giving him extraordinary power in the renegotiations.

This survey of the sovereign bond market provides investors with a helpful toolkit for analyzing sovereign creditworthiness and foreseeing developments in the international financial architecture. The result should be a better understanding of market developments and more deliberate investment decisions. While keeping in mind that credit risk premia compensate for default risk, this monograph highlights the role of the recovery value implied in risk spreads. Awareness of both components is essential for the improvement of investment strategies, especially in the face of financial distress. If this study serves to enhance the functioning and prosperousness of the international market for sovereign debt instruments in the interest of debtors and creditors alike, it will have achieved its goal.

References

Aggarwal, Vinod K. (1996), *Debt games: Strategic interaction in international debt rescheduling*, Cambridge University Press, Cambridge.

Aizenman, Joshua, Kenneth M. Kletzer and Brian Pinto (2005), 'Sargent-Wallace meets Krugman-Flood-Garber, or: Why sovereign debt swaps don't avert macroeconomic crises', *Economic Journal* **115**, 343–367.

Alesina, Alberto and Guido Tabellini (1990), 'A positive theory of fiscal deficits and government debt', *Review of Economic Studies* **57**, 403–414.

Alfaro, Laura and Fabio Kanczuk (2005), 'Sovereign debt as a contingent claim: a quantitative approach', *Journal of International Economics* **65**, 297–314.

Allen, Mark, Christoph Rosenberg, Christian Keller, Brad Setser and Nouriel Roubini (2002), *A balance sheet approach to financial crisis*, IMF Working Paper.

Alles, Lakshman (2001), *Asset securitization and structured financing: future prospects and challenges for emerging market countries*, IMF Working Paper.

Alper, C. Emre, Aras Akdemir and Kazim Kazimov (2004), Estimating the term structure of government securities in Turkey, Working paper, Bogazici University Istanbul.

Altman, Edward I. (1989), 'Measuring corporate bond mortality and performance', *Journal of Finance* **44**, 902–922.

Altman, Edward I., Andrea Resti and Andrea Sironi (2005), *Recovery Risk: The next challenge in credit risk management*, Risk Books, London.

Altman, Edward I., Brooks Brady, Andrea Resti and Andrea Sironi (2002), The link between default and recovery rates: implications for credit risk models and procyclicality, Working paper, New York University.

Altman, Edward I. and Vellore M. Kishore (1996), 'Almost everything you wanted to know about recoveries on defaulted bonds', *Financial Analyst's Journal* **52**, 57–64.

Amato, Jeffrey D. and Eli M. Remolona (2003), *The credit spread puzzle*, BIS Working Paper.

Ammann, Manuel (1999), *Pricing Derivative Credit Risk*, Springer Verlag, Berlin, Heidelberg, New York.

Andritzky, Jochen R. (2004*a*), Analyzing the default risk of sovereign Eurobonds: An application of Kalman filtering on Russian and Turkish bonds, Working paper, Universität St. Gallen.

Andritzky, Jochen R. (2004*b*), Holdouts und Anlegerklagen nach einer Restrukturierung von Staatsschulden, Working paper, University of St. Gallen.

Andritzky, Jochen R. (2005), 'Default and recovery rates of sovereign bonds: a case study of the Argentine crisis', *Journal of Fixed Income* **7**, 97–107.

Andritzky, Jochen R., Geoffrey J. Bannister and Natalia T. Tamirisa (2005), *The impact of macroeconomic announcements on emerging market bonds*, IMF Working Paper.

Ang, Andrew and Monika Piazzesi (2003), 'A no-arbitrage vector autoregression of term structure dynamics with macroeconomic and latent variables', *Journal of Monetary Economics* **50**, 745–787.

Arora, Vivek and Martin Cerisola (2001), *How does U.S. monetary policy influence sovereign spreads in emerging markets*, IMF Working Paper.

Atkeson, Andrew (1991), 'International lending with moral hazard and risk of repudiation', *Econometrica* **59**, 1069–1090.

Aylward, Lynn and Rupert Thorne (1998), *An econometric analysis of countries' repayment performance to the International Monetary Fund*, IMF Working Paper.

Baig, Taimur and Ilan Goldfajn (1999), 'Financial market contagion in the asin crises', *IMF Staff Papers* **46**, 167–195.

Bakshi, Gurdip, Dilip B. Madan and Frank X. Zhang (2001), Understanding the role of recovery in default risk models: Empirical comparisons and implied recovery rates, Working paper, University of Maryland.

Balduzzi, Pierluigi, Sanjiv Ranjan Das, Silverio Foresi and Rangarajan Sundaram (1996), 'A simple approach to three factor affine term structure models', *Journal of Fixed Income* **6**, 43–53.

Balkan, Erol (1992), *Insuring sovereign debt against default*, World Bank Discussion Paper.

Barnett, Barry C., Sergio J. Galvis and Ghislain, jr. Gouraige (1984), 'On Third World debt', *Harvard International Law Journal* **25**, 83–151.

Barro, Robert J. (1979), 'On the determination of public debt', *Journal of Policital Economy* **87**, 940–971.

Barro, Robert J. (1995), *Optimal debt management*, NBER Working Paper.

Barro, Robert J. (2001), *Economic growth in east Asia before and after the financial crisis*, NBER Working Paper.

Barro, Robert J. and Jong-Wha Lee (2001), *IMF programs: Who is chosen and what are the effects?*, IMF Working Paper.

Becker, Torbjörn, Anthony Richards and Yunyong Thaicharoen (2003), 'Bond restructuring and moral hazard: Are collective action clauses costly?', *International Finance* **6**, 415–447.

Bekaert, Geert and Campbell R. Harvey (2002), 'Research in emerging markets finance: looking to the future', *Emerging Markets Review* **3**, 429–448.

Bekaert, Geert and Campbell R. Harvey (2003), 'Emerging markets finance', *Journal of Empirical Finance* **10**, 3–55.

Berardi, Andrea and Michele Trova (2003), The term structure of default probabilities implicit in emerging market bond prices, Working paper, University of Verona.

Berardi, Andrea, Stefania Ciraolo and Michele Trova (2004), 'Predicting default probabilities and implementing trading strategies for emerging market bond portfolios', *Emerging Markets Review* **5**, 447–469.

Berg, Andrew and Catherine Pattillo (1999), 'Predicting currency crises: the indicators approach and an alternative', *Journal of International Money and Finance* **18**, 561–586.

Berg, Andrew, Eduardo Borensztein, Gian Maria Milesi-Ferretti and Catherine Pattillo (1999), *Anticipating Balance of Payments Crises: The Role of Early Warning Systems*, IMF Occasional Paper.

Berg, Andrew and Jeffrey Sachs (1988), 'The debt crisis', *Journal of Development Economics* **29**, 271–306.

Bhanot, Karan (1998), 'Recovery and implied default in Brady bonds', *Journal of Fixed Income* **8**, 47–51.

Bhatia, Ashok Vir (2002), *Sovereign credit rating methodology: an evaluation*, IMF Working Paper.

Bielecki, Tomasz R. and Marek Rutkowski (2002), *Credit Risk: Modeling, Valuation and Hedging*, Springer Verlag, Berlin, Heidelberg, New York.

Bird, Graham and Dane Rowlands (1997), 'The catalytic effect of lending by the international financial institutions', *World Economy* **20**, 967–991.

Bird, Graham and Dane Rowlands (2000), 'The catalyzing role of policy-based lending by the IMF and the World Bank: fact or fiction', *Journal of International Development* **12**, 951–973.

Bird, Graham, Mumtaz Hussain and Joseph P. Joyce (2004), 'Many happy returns? Recidivism and the IMF', *Journal of International Money and Finance* **23**, 231–251.

Bissoondoyal-Bheenick, Emawtee (2005), 'An analysis of the determinants of sovereign ratings', *Global Financial Journal* **15**, 251–280.

Black, Fisher and John C. Cox (1976), 'Valuing corporate securities: Some effects of bond indenture provisions', *Journal of Finance* **31**, 351–367.

Black, Fisher and Myron Scholes (1972), 'The pricing of options and corporate liabilities', *Journal of Political Economy* **81**, 637–654.

Blanco, Roberto, Simon Brennan and Ian W. Marsh (2005), 'An empirical analysis of the dynamic relationships between investment grade bonds and credit default swaps', *Journal of Finance* **60**, 2255–2281.

Bliss, Robert R. (1997), 'Testing term structure estimation methods', *Advances in Futures and Options Research* **9**, 197–231.

Block, Steven A., Burkhard N. Schrage and Paul M. Vaaler (2003), Democratization's risk premium: Partisan and opportunistic political business cycle effects on sovereign ratings in developing countries, Working paper, University of Michigan.

Block, Steven A. and Paul M. Vaaler (2004), 'The price of democracy: Sovereign risk ratings, bond spreads, and political business cycles in developing countries', *Journal of International Money and Finance* **23**, 917–946.

Boehmer, E. and W. Megginson (1990), 'Determinants of secondary market prices for developing country syndicated loans', *Journal of Finance* **45**, 1517–1540.

Bordo, Michael D., Barry Eichengreen and Jongwoo Kim (1998), *Was there really an earlier period of international financial integration comparable to today?*, NBER Working Paper.

Borensztein, Eduardo, Marcos Chamon, Olivier Jeanne, Paolo Mauro and Jeromin Zettelmeyer (2004), *Sovereign debt structure for crisis prevention*, IMF Occasional Paper No. 237.

Bratton, William W. and Gaurang Mitu Gulati (2004), 'Sovereign debt reform and the best interest of creditors', *Vanderbilt Law Review* **57**, 1–72.

Brennan, Michael J. and Eduardo S. Schwartz (1977), 'Convertible bonds: Valuation and optimal strategies for call and conversion', *Journal of Finance* **32**, 1699–1715.

Brennan, Michael J. and Eduardo S. Schwartz (1980), 'Analyzing convertible bonds', *Journal of Financial and Quantitative Analysis* **15**, 907–929.

Brennan, Michael J. and Eduardo S. Schwartz (1982), 'An equilibrium model of bond pricing and a test of market efficiency', *Journal of Financial and Quantitative Analysis* **17**, 301–329.

Brennan, Michael J., Eduardo S. Schwartz and Manoj K. Singh (1979), 'A continuous time approach to the pricing of bonds', *Journal of Banking and Finance* **3**, 133–155.

Bryant, John (1980), 'A model of reserves, bank runs, and deposit insurance', *Journal of Banking and Finance* **4**, 333–344.

Buchheit, Lee C. and Gaurang Mitu Gulati (2000), Exit consents in sovereign bond exchanges, Working paper, University of California, Los Angeles.

Buckley, Ross P. (2004), 'Turning loans into bonds: lessons for East Asia from the Latin American Brady Plan', *Journal of Restructuring Finance* **1**, 185–200.

Bulow, Jeremy and Kenneth Rogoff (1989*a*), 'A constant recontracting model of sovereign debt', *Journal of Political Economy* **97**, 155–178.

Bulow, Jeremy and Kenneth Rogoff (1989*b*), 'Sovereign debt: Is to forgive to forget?', *American Economic Review* **79**, 43–50.

Bulow, Jeremy and Kenneth Rogoff (1991), 'Sovereign debt repurchases: no cure for overhang', *Quarterly Journal of Economics* **106**, 1219–1235.

Callier, Philippe (1985), 'Further results on countries' debt-service performance: the relevance of structural factors', *Weltwirtschaftliches Archiv* **121**, 105–115.

Calvo, Guillermo A. (2000), Balance of payment crises in emerging markets: large capital inflows and sovereign governments, *in* P.Krugman, ed., 'Currency Crises', University Chicago Press, Chicago.

Calvo, Guillermo A. and Enrique G. Mendoza (1999), *Regional contagion and the globalization of securities markets*, NBER Working Paper.

Calvo, Guillermo A., Leonardo Leiderman and Carmen Reinhart (1993), 'Capital inflows and real exchange rate appreciation in Latin America: the role of external factors', *IMF Staff Papers* **40**, 108–151.

Campbell, John Y. and Robert J. Shiller (1991), 'Yield spreads and interest rate movements: A bird's eye view', *Review of Economic Studies* **58**, 495–514.

Cantor, Richard and Frank Packer (1996), 'Determinants and impact of sovereign credit ratings', *FRBNY Economic Policy Review* **2**, 37–54.

Catao, Luis A. and Bennett W. Sutton (2002), *Sovereign defaults: the role of volatility*, IMF Working Paper.

Cerra, Valerie and Chaman Saxena (2005), 'Did output recover from the Asian Crisis', *IMF Staff Papers* **52**, 1–23.

Chan-Lau, Jorge A. and Yoon Sook Kim (2004), *Equity prices, credit default swaps, and bond spreads in emerging markets*, IMF Working Paper.

Chang, Roberto and Andrés Velasco (1999), *Liquidity crises in emerging markets: theory and policy*, NBER Working Paper.

Chang, Roberto and Andrés Velasco (2000), 'Servicing the public debt: the role of expectations', *Journal of International Economics* **1**, 647–661.

Chang, Roberto and Andrés Velasco (2001), 'A model of financial crises in emerging markets', *Quarterly Journal of Economics* **116**, 489–517.

Chari, V. V. and Patrick J. Kehoe (1993), 'Sustainable plans and mutual default', *Review of Economic Studies* **60**, 175–195.

Chen, Nai-fu (1991), 'Financial investment opportunitites and the macroeconomy', *Journal of Finance* **46**, 529–554.

Chuhan, Punam and Federico Sturzenegger (2003), Default episodes in the 1980s and 1990s: What have we learned?, Working paper, Universidad Torcuato Di Tella.

Citron, Joel-Thomas and Gerald Nickelsburg (1987), 'Country risk and political instability', *Journal of Development Economics* **25**, 385–392.

Claessens, Stijn and George Pennacchi (1996), 'Estimating the likelihood of Mexican default from the market prices of Brady bonds', *Journal of Financial and Quantitative Analysis* **31**, 109–126.

Claessens, Stijn and Kristin Forbes (2001), *International Financial Contagion*, Kluwer Academic Publishers, Boston, Dordrecht, London.

Cline, William R. (1984), *International debt: Systemtic risk and policy response*, Institute for International Economics, Washington, DC.

Cline, William R. (2003), *Restoring economic growth in Argentina*, World Bank Policy Research Working Paper.

Cline, William R. (2004), Private sector involvement in financial crisis resolution: definition, measurement and implementation, *in* A.Haldane, ed., 'Fixing Financial Crises in the Twenty-First Century', Routledge, London.

Cline, William R. and Kevin J. S. Barnes (1997), Spreads and risk in emerging market lending, Working paper, Institute of International Finance.

Cohen, Benjamin J. (1989*a*), *Developing country debt: a middle way*, Princeton Essays in International Finance 173, Princeton.

Cohen, Benjamin J. (1989*b*), 'A global Chapter 11', *Foreign Policy* **75**, 109–111.

Cohen, Daniel (1992), *The debt crisis: a post mortem*, CEPR Working Paper.

Cole, Harold L., James Dow and William B. English (1995), 'Default, settlement and signalling: Lending resumption in a reputational model of sovereign debt', *International Economic Review* **36**, 365–385.

Cole, Harold L. and Patrick J. Kehoe (1996*a*), *Reputation spillover across relationships: Reviving reputation models of debt*, NBER Working Paper.

Cole, Harold L. and Timothy J. Kehoe (1996*b*), 'A self-fulfilling model of Mexico's 1994-95 debt crisis', *Journal of International Economics* **41**, 309–330.

Cole, Harold L. and Timothy J. Kehoe (2000), 'Self-fulfilling debt crises', *Review of Economic Studies* **67**, 91–116.

Collin-Dufresne, Pierre and Robert S. Goldstein (2001), 'Do credit spreads reflect stationary leverage ratios?', *Journal of Finance* **56**, 1929–1957.

Collin-Dufresne, Pierre, Robert S. Goldstein and J. Spencer Martin (2001), 'The determinants of credit spread changes', *Journal of Finance* **56**, 2177–2207.

Collin-Dufresne, Pierre, Robert S. Goldstein and Jean Helwege (2003), Is credit event risk priced? Modeling contagion via the updating of beliefs, Working paper, Carnegie Mellon University.

Conklin, James (1998), 'The theory of sovereign debt and Spain under Philip II', *Journal of Policital Economy* **106**, 483–513.

Cornell, Bradford and Kevin Green (1991), 'The investment performance of low-grade bond funds', *Journal of Finance* **46**, 29–48.

Corsetti, Giancarlo, Bernardo Guimaraes and Nouriel Roubini (2005), *International lending of last resort and moral hazard—a model of IMF's catalytic finance*, CEPR Working Paper.

Corsetti, Giancarlo and Nouriel Roubini (1997), 'Politically motivated fiscal deficits: policy issues in closed and open economies', *Economics and Politics* **9**, 27–54.

Corsetti, Giancarlo, Paolo Pesenti and Nouriel Roubini (1999), 'Paper tigers? A model of the Asian crisis', *European Economic Review* **43**, 1211–1236.

Corsetti, Giancarlo, Paolo Pesenti, Nouriel Roubini and Cedric Tille (2000), 'Trade and contagious devaluations: A welfare-based approach', *Journal of International Economics* **50**, 217–241.

Cossin, Didier and Hugues Pirotte (2001), *Advanced credit risk analysis: Financial approaches and mathematical models to asses, price, and manage credit risk*, John Wiley & Sons, New York et al.

Cossin, Didier, Thomas Hricko, Daniel Aunon-Nerin and Zhijiang Huang (2002), Exploring for the determinants of credit risk in credit default swap transaction data: Is fixed-income markets' information sufficient to evaluate credit risk?, Working paper, HEC.

Cottarelli, Carlo and Curzio Giannini (2002), *Bedfellows, hostages, or perfect strangers? Global capital markets and the catalytic effect of IMF crisis lending*, IMF Working Paper.

Cox, John C., Jonathan E. Ingersoll and Stephen A. Ross (1985a), 'An intertemporal general equilibrium model of asset prices', *Econometrica* **53**, 363–384.

Cox, John C., Jonathan E. Ingersoll and Stephen A. Ross (1985b), 'A theory of the term structure of interest rates', *Econometrica* **53**, 385–407.

Cunningham, Alastair, Liz Dixon and Simon Hayes (2001), Analysing yield spreads on emerging market sovereign bonds, Working paper, Bank of England.

Dai, Qiang and Kenneth J. Singleton (2000), 'Specification analysis of affine term structure models', *Journal of Finance* **55**, 1943–1978.

Dailami, Mansoor and Robert Hauswald (2003), *The emerging project bond market: Covenant provisions and credit spreads*, World Bank Policy Research Working Paper.

Das, Sanjiv Ranjan and Peter Tufano (1996), 'Pricing credit sensitive debt when interest rates, credit ratings, and credit spreads are stochastic', *Journal of Financial Engineering* **5**, 161–198.

Daseking, Christina, Atish R. Ghosh, Alun H. Thomas and Timothy Lane (2005), *Lessons from the crisis in Argentina*, IMF Occasional Paper No. 236.

De la Torre, Augusto, Eduardo Levy Yeyati and Sergio Schmukler (2002), Argentina's financial crisis: Floating money, sinking banking, Working paper, World Bank.

Dell'Ariccia, Giovanni, Isabel Schnabel and Jeromin Zettelmeyer (2002), *Moral hazard and international crisis lending: a test*, IMF Working Paper.

Detragiache, Enrica (1996), *Rational liquidity crises in the sovereign debt market: in search of a theory*, IMF Working Paper.

Detragiache, Enrica and Antonio Spilimbergo (2001), *Crisis and liquidity: evidence and interpretation*, IMF Working Paper.

Diamond, Douglas and Philip H. Dybvig (1983), 'Bank runs, deposit insurance, and liquidity', *Journal of Political Economy* **91**, 401–419.

Dicks-Mireaux, Louis, Mauro Mecagni and Susan Schadler (2000), 'Evaluating the effects of IMF lending to low-income countries', *Journal of Development Economics* **61**, 495–526.

Diebold, Francis X., Monika Piazzesi and Glenn D. Rudebusch (2005), Modeling bond yields in finance and macroeconomics, Working paper, University of Chicago.

Dixon, Liz and David Wall (2000), 'Collective action problems and collective action clauses', *Financial Stability Review* **9**, 142–151.

Dooley, Michael P. (1986), *An analysis of the debt crisis*, IMF Working Paper.

Dooley, Michael P. (1988), 'Buy-backs and market valuation of external debt', *IMF Staff Papers* **35**, 215–229.

Dooley, Michael P. (2000), *Can output losses following international financial crises be avoided?*, NBER Working Paper.

Dooley, Michael P., Eduardo Fernandez-Arias and Kenneth M. Kletzer (1994), *Recent private capital inflows to developing countries: Is the debt crisis history?*, NBER Working Paper.

Dooley, Michael P. and Jeffrey A. Frankel (2003), *Managing currency crises in emerging markets*, NBER Working Paper.

Dooley, Michael P. and Sujata Verma (2001), *Rescue packages and output losses following crises*, NBER Working Paper.

Drazen, Allan and Paul R. Masson (1994), 'Credibility of policies versus credibility of policymakers', *Quarterly Journal of Economics* **109**, 735–754.

Dreher, Axel and Roland Vaubel (2004), 'Do IMF and IBRD cause moral hazard and political business cycles? Evidence from panel data', *Open Economies Review* **15**, 5–22.

Duffee, Gregory R. (1996), 'Idiosyncratic variation of Treasury bill yields', *Journal of Finance* **51**, 527–551.

Duffee, Gregory R. (1998), 'The relation between Treasury yields and corporate bond yield spreads', *Journal of Finance* **53**, 2225–2241.

Duffee, Gregory R. (2002), 'Term premia and interest rate forecasts in affine models', *Journal of Finance* **57**, 405–443.

Duffee, Gregory R. and Richard Stanton (2004), Estimation of dynamic term structure models, Working paper, University of California, Berkeley.

Duffie, Darrell (1999*a*), 'Credit swap valuation', *Financial Analyst's Journal* **55**, 73–87.

Duffie, Darrell (1999*b*), Defaultable term structure models with fractional recovery of par, Working paper, Stanford University.

Duffie, Darrell and Kenneth J. Singleton (1997), 'An econometric model of the term structure of interest-rate swap yields', *Journal of Finance* **52**, 1287–1321.

Duffie, Darrell and Kenneth J. Singleton (1999), 'Modeling term structures of defaultable bonds', *The Review of Financial Studies* **12**, 687–720.

Duffie, Darrell, Lasse Heje Pedersen and Kenneth J. Singleton (2003), 'Modeling sovereign yield spreads: A case study of Russian debt', *Journal of Finance* **58**, 119–159.

Duffie, Darrell and Ming Huang (1996), 'Swap rates and credit quality', *Journal of Finance* **51**, 921–949.

Duffie, Darrell and Rui Kan (1996), 'A yield-factor model of interest rates', *Mathematical Finance* **6**, 379–406.

Eaton, Jonathan (1993), 'Sovereign debt: A primer', *World Bank Economic Review* **7**, 137–172.

Eaton, Jonathan (2002), Standstills and an international bankruptcy court, Working paper, New York University.

Eaton, Jonathan and Mark Gersovitz (1981), 'Debt with potential repudiation: Theoretical and empirical analysis', *Review of Economic Studies* **48**, 289–309.

Eaton, Jonathan, Mark Gersovitz and Joseph E. Stiglitz (1986), 'The pure theory of country risk', *European Economic Review* **30**, 481–513.

Eaton, Jonathan and Raquel Fernandez (1995), *Sovereign debt*, NBER Working Paper.

Edwards, Sebastian (1986), 'The pricing of bonds and bank loans in international markets: An empirical analysis of developing countries' foreign borrowing', *European Economic Review* **30**, 565–589.

Edwards, Sebastian (1989), *The International Monetary Fund and the developing countries: a critical evaluation*, NBER Working Paper.

Edwards, Sebastian and Jeffrey A. Frankel (2002), *Preventing Currency Crises in Emerging Markets*, Cambridge University Press, Cambridge.

Eichengreen, Barry and Ashoka Mody (1998a), 'Interest rates in the North and capital flows to the South: Is there a missing link?', *International Finance* **1**, 35–57.

Eichengreen, Barry and Ashoka Mody (1998b), *What explains changing spreads on emerging-market debt: Fundamentals or market sentiment*, NBER Working Paper.

Eichengreen, Barry and Ashoka Mody (2000a), *Bail-ins, bailouts, and borrowing costs*, IMF Working Paper.

Eichengreen, Barry and Ashoka Mody (2000b), *Would collective action clauses raise borrowing costs?*, NBER Working Paper.

Eichengreen, Barry and Ashoka Mody (2000c), Would collective action clauses raise borrowing costs? An update and additional results, Working paper, University of California, Berkeley.

Eichengreen, Barry and Christof Rühl (2000), *The bail-in problem: Systematic goals, ad hoc means*, NBER Working Paper.

Eichengreen, Barry, Kenneth M. Kletzer and Ashoka Mody (2003), *Crisis resolution: Next steps*, NBER Working Paper.

Eichengreen, Barry, Ricardo Hausmann and Ugo Panizza (2003), The pain of original sin, Working paper, University of California, Berkeley.

Eichengreen, Barry and Richard Portes (1986), 'Debt and default in the 1930s: Causes and consequences', *European Economic Review* **30**, 559–640.

Eichengreen, Barry and Richard Portes (1989a), *Dealing with debt: The 1930s and the 1980s*, NBER Working Paper.

Eichengreen, Barry and Richard Portes (1989b), 'Settling defaults in the era of bond finance', *World Bank Economic Review* **3**, 211–239.

Eichengreen, Barry and Richard Portes (1995), *Crisis? What crisis? Orderly workouts for sovereign debtors*, Centre for Economic Policy Research, London.

English, William B. (1996), 'Understanding the cost of sovereign default: American state debt in the 1840's', *American Economic Review* **86**, 259–275.

Erb, Claude B., Campbell R. Harvey and Tadas E. Viskanta (1996), 'Political risk, economic risk, and financial risk', *Financial Analyst's Journal* **52**, 29–46.

Erb, Claude B., Campbell R. Harvey and Tadas E. Viskanta (1999), 'A new perspective of emerging market bonds', *Journal of Portfolio Management* **25**, 83–92.

Fama, Eugene (1970), 'Efficient capital markets: A review of theory and empirical work', *Journal of Finance* **25**, 383–417.

Fama, Eugene (1975), 'Short-term interest rates as predictors of inflation', *American Economic Review* **65**, 269–282.

Fama, Eugene and Kenneth R. French (1993), 'Common risk factors in the returns on stocks and bonds', *Journal of Financial Economics* **33**, 3–56.

Feldstein, Martin S. (2002), *Economic and Financial Crises in Emerging Market Economies*, University Chicago Press, Chicago.

Ferrucci, Gianluigi (2003), Empirical determinants of emerging market economics' sovereign bond spreads, Working paper, Bank of England.

Fisch, Jill E. and Caroline M. Gentile (2004), Vultures or vanguards? The role of litigation in sovereign debt restructuring, Working paper, Fordham University School of Law.

Fishlow, Albert (1985), 'Lessons from the past: Capital markets during the 19th century and the interwar period', *International Organization* **39**, 383–439.

Flood, Robert and Peter Garber (1984), 'Collapsing exchange rate regimes: some linear examples', *Journal of International Economics* **17**, 1–13.

Frank Jr., Charles R. and William R. Cline (1971), 'Measurement of debt servicing capacity: An application of discriminant analysis', *Journal of International Economics* **1**, 327–344.

Franks, Julian and Walter N. Torous (1994), 'A comprison of financial recontracting in distressed exchanges and Chapter 11 reorganizations', *Journal of Financial Economics* **35**, 349–370.

Friedmann, Irving (1977), *The emerging role of private banks in the developing world*, Citicorp, New York.

Frye, Jon (2000a), Collateral damage detected, Working paper, Federal Reserve Bank of Chicago.

Frye, Jon (2000b), Depressing recoveries, Working paper, Federal Reserve Bank of Chicago.

Gai, Prasanna and Ashley Taylor (2004), 'International financial rescues and debtor-country moral hazard', *International Finance* **7**, 391–420.

Galloway, Jennifer (2003), 'What CACs lack', *LatinFinance* **17**, 24–27.

Gande, Amar and David C. Parsley (2005), 'News spillovers in the sovereign debt market', *Journal of Financial Economics* **75**, 691–734.

Geske, Robert (1977), 'The valuation of corporate liabilities as compound options', *Journal of Financial and Quantitative Analysis* **12**, 541–552.

Ghosal, Sayatan and Marcus Miller (2003), 'Co-ordination failure, moral hazard and sovereign bankruptcy procedures', *Economic Journal* **113**, 276–304.

Giannini, Curzio (1999), *Enemy of none but a common friend of all? An international perspective on the lender-of-last-resort function*, IMF Working Paper.

Gibson, Rajna and Suresh M. Sundaresan (1999), A model of sovereign borrowing and sovereign yield spreads, Working paper, HEC.

Goodhart, Charles A. E. and Haizhou Huang (2000), *A simple model of an international lender of last resort*, IMF Working Paper.

Grossmann, Herschel I. and John B. van Huyck (1988), 'Sovereign debt as a contingent claim: Excusable default, repudiation and reputation', *American Economic Review* **78**, 1088–1097.

Group of Ten (1996), *The resolution of sovereign liquidity crisis*, Washington, DC.

Gugiatti, Mark and Anthony Richards (2003), Do collective action clauses influence bond yields? New evidence from emerging markets, Discussion paper, Reserve Bank of Australia.

Gulati, Gaurang Mitu and Kenneth N. Klee (2001), 'Sovereign piracy', *Business Lawyer* **56**, 635–651.

Hajivassiliou, Vassilis A. (1987), 'The external debt repayment problems of LDC's: an econometric model based on panel data', *Journal of Econometrics* **36**, 205–230.

Hajivassiliou, Vassilis A. (1989), *Do the secondary markets believe in life after debt*, Cowles Foundation Discussion Papers.

Hajivassiliou, Vassilis A. (1994), 'A simulation estimation analysis of the external debt crises of developing countries', *Journal of Applied Econometrics* **9**, 109–113.

Haldane, Andrew G. (1999), *Private sector involvement in financial crises: analytics and public policy approaches*, IMF Working Paper.

Haldane, Andrew G., Adrian Penalver, Victoria Saporta and Hyun Song Shin (2005), 'Analytics of sovereign debt restructuring', *Journal of International Economics* **65**, 315–333.

Haque, Nadeem U. and Mohsin S. Khan (1998), *Do IMF-supported programs work? A survey of the cross-country empirical evidence*, IMF Working Paper.

Haque, Nadeem U., Nelson Mark and Donald J. Mathieson (1998), *The relative importance of political and economic variables in creditworthiness ratings*, IMF Working Paper.

Heath, David C., Robert A. Jarrow and Andrew J. Morton (1992), 'Bond pricing and the term structure of interest rates: A new methodology for contingent claims valuation', *Econometrica* **60**, 77–105.

Helpman, Elhanan (1989), Voluntary debt reduction: incentives and welfare, *in* J.Frenkel, M.Dooley and P.Wickham, eds, 'Analytical issues in debt', International Monetary Fund, Washington, DC.

Ho, Thomas S. Y. and Sang-Bin Lee (1986), 'Term structure movements and pricing interest rate claims', *Journal of Finance* **41**, 1011–1029.

Hobbes, Thomas (1968), *Leviathan*, Penguin, Harmondsworth.

Hu, Yen-Ting, Rudiger Kiesel and William Perraudin (2002), 'The estimation of transition matrices for sovereign credit ratings', *Journal of Banking and Finance* **26**, 1383–1406.

Hu, Yen-Ting and William Perraudin (2002), *The dependence of recovery rates and default*, CEPR Working Paper.

Hull, John C. and Alan White (1990), 'Pricing of interest-rate-derivative securities', *The Review of Financial Studies* **3**, 573–592.

Hull, John C. and Alan White (1994), 'Numerical procedures for implementing term structure models II: Two-factor models', *Journal of Derivatives* **2**, 37–48.

Hull, John C. and Alan White (2000), 'Valuing credit default swaps I: No counterparty default risk', *The Journal of Derivatives* **8**, 29–40.

Hull, John C. and Alan White (2001), 'Valuing credit default swaps II: Modeling default correlations', *The Journal of Derivatives* **8**, 12–22.

Hull, John C., Mirela Predescu and Alan White (2004), 'The relationship between credit default swap spreads, bond yields, and credit rating announcements', *Journal of Banking and Finance* **28**, 2789–2811.

Hull, John C., Mirela Predescu and Alan White (2005), 'Bond prices, default probabilities and risk premiums', *Journal of Credit Risk* **1**, 53–60.

Hutchison, Michael M. (2001), *A cure worse than the disease? Currency crises and the output costs of IMF-supported stabilization programs*, NBER Working Paper.

International Monetary Fund (2001), *Involving the private sector in the resolution of financial crisis—restructuring international sovereign bonds*, Washington, DC.

International Monetary Fund (2002), *Fund policy on lending into arrears to private creditors—further consideration of the Good Faith criterion*, Washington, DC.

International Monetary Fund (2003), *Reviewing the process for sovereign debt restructuring within the existing legal framework*, Washington, DC.

International Monetary Fund (2005), *Progress report to the International Monetary and Financial Committee on crisis resolution*, Washington, DC.

Izvorski, Ivailo (1998), *Brady bonds and default probabilities*, IMF Working Paper.

Jarrow, Robert A. (2001), 'Default parameter estimation using market prices', *Financial Analyst's Journal* **57**, 75–92.

Jarrow, Robert A. and Stuart M. Turnbull (1995), 'Pricing derivatives on financial securities subject to credit risk', *Journal of Finance* **50**, 53–85.

Jeanne, Olivier (2000), 'Foreign currency debt and the global financial architecture', *European Economic Review* **44**, 719–727.

Jeanne, Olivier and Charles Wyplosz (2001), *The international lender of last resort: How large is large enough?*, NBER Working Paper.

Jeanne, Olivier and Jeromin Zettelmeyer (2001), 'International bailouts, moral hazard, and conditionality', *Economic Policy* **33**, 409–432.

Jeanne, Olivier and Jeromin Zettelmeyer (2004), *The Mussa Theorem (and other results on IMF-induced moral hazard)*, IMF Working Paper.

Jostova, Gergana (2006), 'Predictability in emerging sovereign debt markets', *Journal of Business* **79**.

Kamin, Steven B. and Karsten von Kleist (1999), *The evolution and determinants of emerging market credit spreads in the 1990s*, BIS Working Paper.

Kaminsky, Graciela L. and Carmen Reinhart (1999), 'The twin crises: The cause of banking and balance-of-payment problems', *American Economic Review* **89**, 473–500.

Karmann, Alexander and Dominik Maltritz (2002), Sovereign risk in a structural approach—evaluating sovereign ability-to-pay and probability of default, Working paper, Dresden University of Technology.

Kehoe, Patrick J. and David K. Levine (1993), 'Debt-constrained asset markets', *Review of Economic Studies* **60**, 865–888.

Kenen, Peter B. (1998), *Should the IMF pursue capital-account convertibility*, Princeton Essays in International Finance 207, Princeton.

Ketkar, Suhas and Dilip Ratha (2001), 'Securitization of future flow receivables: A useful tool for developing countries', *Finance & Development* **38**, 46–49.

Kharas, Homi, Brian Pinto and Sergei Ulatov (2001), 'An analysis of Russia's 1998 meltdown: Fundamentals and market signals', *Brookings Papers on Economic Activity* **2001**, 1–50.

Killick, Tony (1997), 'Principals, agents, and the failing of conditionality', *Journal of International Development* **9**, 483–495.

Kletzer, Kenneth M. (2003), Sovereign bond restructuring: Collective action clauses and official crisis intervention, Working paper, University of California, Santa Cruz.

Kletzer, Kenneth M. and Brian D. Wright (2000), 'Sovereign debt as an intertemporal barter', *American Economic Review* **90**, 621–639.

Klingen, Christoph, Beatrice Weder and Jeromin Zettelmeyer (2004), *How private creditors fared in emerging debt markets, 1970–2000*, IMF Working Paper.

Krueger, Anne (2001), *International Financial Architecture for 2002: A New Approach to Sovereign Debt Restructuring*, International Monetary Fund, Washington, DC.

Krugman, Paul (1979), 'A model of balance of payment crises', *Journal of Money, Credit, and Banking* **11**, 311–325.

Krugman, Paul (1988), 'Financing vs. forgiving a debt overhang', *Journal of Development Economics* **29**, 253–268.

Kulatilaka, N. and A. J. Marcus (1987), 'A model of strategic default of sovereign debt', *Journal of Economic Dynamics and Control* **11**, 483–498.

Kumar, Manmohan S., Paul R. Masson and Marcus Miller (2000), *Global financial crises: Institutions and incentives*, IMF Working Paper.

Lando, David (1998), 'On Cox processes and credit risky securities', *Review of Derivatives Research* **2**, 99–120.

Lane, Timothy and Steven Phillips (2000), *Does IMF financing result in moral hazard*, IMF Working Paper.

Lanoie, Paul and Sylvain Lemarbre (1996), 'Three approaches to predict the timing and quantity of LDC debt rescheduling', *Economics Letters* **28**, 241–246.

Larrain, Guillermo, Helmut Reisen and Julia von Maltzan (1997), Emerging market risk and sovereign credit ratings, Working paper, OECD Development Center.

Lee, Suk Hun (1991), 'Ability and willingness to service debt as explanation for commercial and official rescheduling cases', *Journal of Banking and Finance* **15**, 5–27.

Leland, Hayne E. (1994), 'Corporate debt value, bond covenants, and optimal capital structure', *Journal of Finance* **49**, 1213–1252.

Leland, Hayne E. and Klaus Bjerre Toft (1996), 'Optimal capital structure, endogenous bankruptcy, and the term structure of credit spreads', *Journal of Finance* **51**, 987–1019.

Lerrick, Adam and Allan H. Meltzer (2001), Blueprint for an international lender of last resort, Working paper, Carnegie Mellon University.

Li, Carmen A. (1992), 'Debt arrears in Latin America: do political variables matter', *Journal of Development Studies* **28**, 668–688.

Lindert, Peter H. and Peter J. Morton (1989), How sovereign debt has worked, *in* J.Sachs, ed., 'Developing Country Debt and the World Economy', University Chicago Press, Chicago.

Lipworth, Gabrielle and Jens Nystedt (2001), 'Crisis resolution and private sector adaption', *IMF Staff Papers* **47**, 188–214.

Lloyd-Ellis, Huw, George W. McKenzie and Steve H. Thomas (1989), 'Using country balance sheet data to predict debt rescheduling', *Economics Letters* **31**, 173–177.

Lloyd-Ellis, Huw, George W. McKenzie and Steve H. Thomas (1990), 'Predicting the quantity of LDC debt rescheduling', *Economics Letters* **32**, 67–73.

Longstaff, Francis A. (2004), 'The flight-to-liquidity premium in US Treasury bond prices', *Journal of Business* **77**, 511–526.

Longstaff, Francis A. and Eduardo S. Schwartz (1992), 'Interest rate volatility and the term structure: A two-factor general equilibrium model', *Journal of Finance* **47**, 1259–1282.

Longstaff, Francis A. and Eduardo S. Schwartz (1995), 'A simple approach to valuing risky fixed and floating rate debt', *Journal of Finance* **50**, 789–819.

Longstaff, Francis A., Sanjay Mithal and Eric Neis (2004), *Corporate yield spreads: default risk or liquidity? New evidence from the credit-default swap market*, NBER Working Paper.

Madan, Dilip B. and Haluk Unal (1998), 'Pricing the risks of default', *Review of Derivatives Research* **2**, 121–160.

Manasse, Paolo and Nouriel Roubini (2005), *"Rules of thumb" for sovereign debt crises*, IMF Working Paper.

Manasse, Paolo, Nouriel Roubini and Axel Schimmelpfennig (2003), *Predicting sovereign debt crises*, IMF Working Paper.

Marashaden, Omar (1997), 'A logit model to predict debt rescheduling by less developed countries', *Asian Economies* **26**, 25–34.

Marchesi, Silvia (2003), 'Adoption of an IMF programme and debt rescheduling. An empirical analysis', *Journal of Development Economics* **70**, 403–423.

Mauro, Paolo, Nathan Sussman and Yishay Yafeh (2002), 'Emerging market spreads: Then versus now', *Quarterly Journal of Economics* **117**, 695–733.

Mauro, Paolo and Yishay Yafeh (2003), *The corporation of foreign bondholders*, IMF Working Paper.

McBrady, Matthew R. and Mark S. Seasholes (2004), 'Bailing-in', *Journal of Restructuring Finance* **1**, 49–77.

McCulloch, J. Houston (1971), 'Measuring the term structure of interest rates', *Journal of Business* **44**, 19–31.

McCulloch, J. Houston (1975), 'The tax adjusted yield curve', *Journal of Finance* **30**, 811–830.

McFadden, Daniel, Richard Eckaus, Gershon Feder, Vassilis A. Hajivassiliou and Stephen O'Connell (1985), Is there life after debt? An econometric analysis of the creditworthiness of developing countries, *in* G.Smith and J.Cuddington, eds, 'International Debt and the Developing Countries', The World Bank, Washington, DC.

McGuire, Patrick and Martijn A. Schrijvers (2003), *Common factors in emerging market spreads*, BIS Working Paper.

McNamara, Gerry and Paul M. Vaaler (2000), 'The influence of competitive positioning and rivalry on emerging market risk assessment', *Journal of International Business Studies* **31**, 337–347.

Merrick, John J. (2001), 'Crisis dynamics of implied default recovery ratios: Evidence from Russia and Argentina', *Journal of Banking and Finance* **25**, 1921–1939.

Merrick, John J. (2005), Evaluating pricing signals from the bond markets, *in* J.Aizenman and B.Pinto, eds, 'Managing Volatility and Crisis: A Practioner's Guide', Cambridge University Press, Cambridge.

Merton, Robert C. (1974), 'On the pricing of corporate debt: The risk structure of interest rates', *Journal of Finance* **29**, 449–470.

Merton, Robert C. (1981), 'On market timing and investment performance. I. An equilibrium theory of value for market forecasts', *Journal of Business* **54**, 363–406.

Miller, Marcus and Lei Zhang (2000), 'Sovereign liquidity crises: A strategic case for a payments standstill', *Economic Journal* **110**, 335–362.

Mishkin, Fredric (1990), 'What does the term structure tell us about future inflation?', *Journal of Monetary Economics* **25**, 77–95.

Mody, Ashoka and Diego Saravia (2003), *Catalyzing capital flows: Do IMF-supported programs work as commitment devices?*, IMF Working Paper.

Mussa, Michael (2002), Argentina and the Fund: From triumph to tragedy, Working paper, Institute for International Economics.

Narag, Ratika (2004), The term structure and default risk in emerging markets, Working paper, University of California, Los Angeles.

Nelson, Charles R. and Andrew F. Siegel (1987), 'Parsimonious modeling of yield curves', *Journal of Business* **60**, 473–489.

Norden, Lars and Martin Weber (2004), 'Informational efficiency of credit default swaps and stock markets: The impact of credit rating announcements', *Journal of Banking and Finance* **28**, 2813–2843.

Obstfeld, Maurice (1994), *The logic of currency crises*, NBER Working Paper.

Odedokun, M. O. (1995), 'Analysis of probability of external debt rescheduling in sub-saharan Africa', *Scottish Journal of Political Economy* **42**, 82–98.

Oechsli, Christopher G. (1981), 'Procedural guidelines for renegotiating LDC debts: An analogy to Chapter 11 of the U.S. Bankruptcy Reform Act', *Virginia Journal of International Law* **21**, 305–341.

O'Rourke, Kevin and Jeffrey G. Williamson (1998), *Globalization and History: The Evolution of a Nineteenth Century Atlantic Economy*, MIT Press, Cambridge, MA.

Ozler, Sule (1993), 'Have commercial banks ignored history?', *American Economic Review* **83**, 608–620.

Packer, Frank and Chamaree Suthiphongchai (2003), *Sovereign credit default swaps*, BIS Working Paper.

Packer, Frank and Haibin Zhu (2005), *Contractual terms and CDS pricing*, BIS Working Paper.

Pages, Henri (2001), *Can liquidity risk be subsumed in credit risk? A case study from Brady bond prices*, BIS Working Paper.

Palacios, Luisa (2004), 'Precedents in sovereign bond default and restructuring', *Journal of Restructuring Finance* **1**, 155–171.

Pan, Jun and Kenneth J. Singleton (2005), Default and recovery implicit in the term structure of sovereign CDS spreads, Working paper, Massachusetts Institute of Technology.

Pericoli, Marcello and Massimo Sbracia (2003), 'A primer on financial contagion', *Journal of Economic Surveys* **17**, 571–608.

Perry, Guillermo and Luis Serven (2003), *The anatomy of a multiple crisis: Why was Argentina special and what can we learn from it?*, World Bank Policy Research Working Paper.

Pescatori, Andrea and Amadou N. R. Sy (2004), *Debt crises and the development of international capital markets*, IMF Working Paper.

Peter, Marcel (2002), Estimating default probabilities of emerging market sovereigns: A new look at the not-so-new literature, Working paper, Graduate Institute of International Studies Geneva.

Prezeworski, Adam and James R. Vreeland (2000), 'The effect of IMF programs on economic growth', *Journal of Development Economics* **62**, 385–421.

Raffer, Kunibert (1990), 'Applying Chapter 9 insolvency to international debt: An economically efficient solution with a human face', *World Development* **18**, 301–311.

Reinhart, Carmen (2002), *Default, Currency Crises and Sovereign Credit Ratings*, NBER Working Paper.

Reisen, Helmut and Julia von Maltzan (1999), 'Boom and bust and sovereign ratings', *International Finance* **2**, 273–293.

Renault, Olivier and Olivier Scaillet (2004), 'On the way to recovery: A nonparametric bias free estimation of recovery rate densities', *Journal of Banking and Finance* **28**, 2915–2931.

Rieffel, Alexis (2003), *Restructuring sovereign debt: The case for ad hoc machinery*, The Brookings Institution, Washington, DC.

Rivoli, Pietra and Thomas L. Brewer (1997), 'Political instability and country risk', *Global Financial Journal* **8**, 309–321.

Rocha, Katia and Francisco A. Garcia (2005), 'Term strucutre of sovereign spreads in emerging markets—a calibration approach for structural models', *Journal of Fixed Income* **14**, 45–57.

Rogoff, Kenneth and Jeromin Zettelmeyer (2002), *Early ideas on sovereign bankruptcy reorganization: a survey*, IMF Working Paper.

Rose, Andrew K. (2005), 'One reason countries pay their debts: Renegotiation and international trade', *Journal of Development Economics* **77**, 189–206.

Rosenthal, Robert (1991), 'On the incentives associated with foreign debt', *Journal of International Economics* **30**, 167–176.

Roubini, Nouriel and Brad Setser (2004a), *Bailouts or Bail-Ins?*, Institute for International Economics, Washington, DC.

Roubini, Nouriel and Brad Setser (2004b), 'The reform of the sovereign debt restructuring process: Problems, proposed solutions, and the Argentine episode', *Journal of Restructuring Finance* **1**, 173–184.

Sachs, Jeffrey, Aaron Tornell and Andrés Velasco (1996), 'The Mexican peso crisis: Sudden death or death foretold', *Journal of International Economics* **41**, 265–283.

Sachs, Jeffrey and Andrew Warner (1995), 'Economic reform and the process of global integration', *Brookings Papers on Economic Activity* **1**, 1–118.

Schönbucher, Philipp J. (2003), *Credit Derivatives Pricing Models: Model, Pricing and Implementation*, John Wiley & Sons, New York et al.

Schönbucher, Philipp J. (2004), 'A measure of survival', *RISK* **17**, 79–85.

Schneider, Martin and Aaron Tornell (2004), 'Balance sheet effects, bailout guarantees and financial crises', *Review of Economic Studies* **71**, 883–913.

Schwarcz, Steven L. (2004), 'Subnational debt restructuring and the rule of law', *Journal of Restructuring Finance* **1**, 129–153.

Söderlind, Paul and Lars E. O. Svensson (1997), 'New techniques to extract market expectations from financial instruments', *Journal of Monetary Economics* **40**, 383–429.

Sharpe, William F. (1994), 'The Sharpe ratio', *Journal of Portfolio Management* **21**, 49–58.

Shea, Gary S. (1985), 'Interest rate term structure estimation with exponential splines: a note', *Journal of Finance* **40**, 319–325.

Shleifer, Andrei (2003), *Will the sovereign debt market survive?*, NBER Working Paper.

Singh, Manmohan (2003a), *Are credit default swap spreads high in emerging markets? An alternative methodology for proxying recovery value*, IMF Working Paper.

Singh, Manmohan (2003b), *Recovery rates from distressed debt— empirical evidence from Chapter 11 filings, international litigation, and recent sovereign debt restructurings*, IMF Working Paper.

Spiegel, Mark M. (2005), 'Solvency runs, sunspot runs, and international bailouts', *Journal of International Economics* **65**, 203–219.

Sturzenegger, Federico (2002), Default episodes in the 90s: Factbook and preliminary lessons, Working paper, Universidad Torcuato Di Tella.

Sturzenegger, Federico and Jeromin Zettelmeyer (2005), *Haircuts: Estimating investor losses in sovereign debt restructurings, 1998–2005*, IMF Working Paper.

Subramanian, KV (2001), 'Term structure estimation in illiquid government bond markets: An empirical analysis for India', *Journal of Fixed Income* **3**, 77–86.

Svensson, Lars E. O. (1994), *Estimating and interpreting forward interest rates: Sweden 1992-94*, NBER Working Paper.

Svensson, Lars E. O. (1995), 'Estimating forward interest rates with the extended Nelson-Siegel method', *Sverigs Riksbank Quarterly Review* **3**, 13.

Sy, Amadou N. R. (2002), 'Emerging market bond spreads and sovereign credit ratings: Reconciling market views with economic fundamentals', *Emerging Markets Review* **3**, 308–408.

Sy, Amadou N. R. (2004), 'Rating the rating agencies: Anticipating currency crises or debt crises', *Journal of Banking and Finance* **28**, 2845–2867.

Sylla, Richard and John Joseph Wallis (1998), 'The anatomy of sovereign debt crises: Lessons from the American state defaults of the 1840s', *Japan and the World Economy* **10**, 267–293.

Taylor, John B. (2002), Sovereign debt restructuring: A U.S. perspective—remarks at the conference on sovereign debt workouts: Hopes and hazards, Discussion paper, Institute for International Economics.

Thomas, Jonathan P. (2004), 'Bankruptcy proceedings for sovereign state insolvency and their effect on capital flows', *International Review of Economics & Finance* **13**, 341–361.

Thomas, Jonathan P. and Silvia Marchesi (1999), 'IMF conditionality as a screening device', *Economic Journal* **109**, 111–125.

Unal, Haluk, Dilip B. Madan and Levent Güntay (2003), 'Pricing the risk of recovery in default with absolute priority rule violation', *Journal of Banking and Finance* **27**, 1001–1025.

Uribe, Martin and Vivian Z. Yue (2003), *Country spreads and emerging countries: who drives whom?*, NBER Working Paper.

Vasicek, Oldrich (1977), 'An equilibrium characterization of the term structure', *Journal of Financial Economics* **5**, 177–188.

Vasicek, Oldrich and H. Gifford Fong (1982), 'Term structure modeling using exponential splines', *Journal of Finance* **37**, 339–348.

Weidenmier, Marc D. (2005), 'Gunboats, reputation, and sovereign repayment: Lessons from the Southern Confederacy', *Journal of International Economics* **66**, 407–422.

Weinschelbaum, Federico and José Wynne (2005), 'Renegotiation, collective action clauses and sovereign debt markets', *Journal of International Economics* **67**, 47–72.

Westphalen, Michael (2001), Valuation of sovereign debt with strategic defaulting and rescheduling, Working paper, HEC.

World Bank (2002), *Financing the poorest countries*, Global Development Finance.

Wright, Mark L.J. (2002), Creditor coordination and sovereign risk, Working paper, Stanford University.

Yeh, Shih-Kuo and Bing-Heui Lin (2003), 'Term structure fitting models and information content: An empirical examination in Taiwanese government bond market', *Review of Pacific Basin Financial Markets and Policies* **6**, 305–348.

Zettelmeyer, Jeromin and Priyadarshani Joshi (2005), *Implicit transfers in IMF lending, 1973–2003*, IMF Working Paper.

Zhang, Frank X. (2003), What did the credit market expect of Argentina default? Evidence from default swap data, Working paper, Federal Reserve Board.

Zhang, Xiaoming A. (1999), Testing for moral hazard in emerging markets lending, Working paper, Institute of International Finance.

Zhou, Chunsheng (1997), A jump-diffusion approach to modeling credit risk and valuing defaultable securities, Working paper, Federal Reserve Board.

Özmen, Erdal and Deniz Arinsoy (2005), 'The original sin and the blessing trinity: An investigation', *Journal of Policy Modeling* **27**, 599–609.

Lecture Notes in Economics and Mathematical Systems

For information about Vols. 1–489
please contact your bookseller or Springer-Verlag

Vol. 495: I. Konnov, Combined Relaxation Methods for Variational Inequalities. XI, 181 pages. 2001.

Vol. 496: P. Weiß, Unemployment in Open Economies. XII, 226 pages. 2001.

Vol. 497: J. Inkmann, Conditional Moment Estimation of Nonlinear Equation Systems. VIII, 214 pages. 2001.

Vol. 498: M. Reutter, A Macroeconomic Model of West German Unemployment. X, 125 pages. 2001.

Vol. 499: A. Casajus, Focal Points in Framed Games. XI, 131 pages. 2001.

Vol. 500: F. Nardini, Technical Progress and Economic Growth. XVII, 191 pages. 2001.

Vol. 501: M. Fleischmann, Quantitative Models for Reverse Logistics. XI, 181 pages. 2001.

Vol. 502: N. Hadjisavvas, J. E. Martínez-Legaz, J.-P. Penot (Eds.), Generalized Convexity and Generalized Monotonicity. IX, 410 pages. 2001.

Vol. 503: A. Kirman, J.-B. Zimmermann (Eds.), Economics with Heterogenous Interacting Agents. VII, 343 pages. 2001.

Vol. 504: P.-Y. Moix (Ed.), The Measurement of Market Risk. XI, 272 pages. 2001.

Vol. 505: S. Voß, J. R. Daduna (Eds.), Computer-Aided Scheduling of Public Transport. XI, 466 pages. 2001.

Vol. 506: B. P. Kellerhals, Financial Pricing Models in Con-tinuous Time and Kalman Filtering. XIV, 247 pages. 2001.

Vol. 507: M. Koksalan, S. Zionts, Multiple Criteria Decision Making in the New Millenium. XII, 481 pages. 2001.

Vol. 508: K. Neumann, C. Schwindt, J. Zimmermann, Project Scheduling with Time Windows and Scarce Resources. XI, 335 pages. 2002.

Vol. 509: D. Hornung, Investment, R&D, and Long-Run Growth. XVI, 194 pages. 2002.

Vol. 510: A. S. Tangian, Constructing and Applying Objective Functions. XII, 582 pages. 2002.

Vol. 511: M. Külpmann, Stock Market Overreaction and Fundamental Valuation. IX, 198 pages. 2002.

Vol. 512: W.-B. Zhang, An Economic Theory of Cities.XI, 220 pages. 2002.

Vol. 513: K. Marti, Stochastic Optimization Techniques. VIII, 364 pages. 2002.

Vol. 514: S. Wang, Y. Xia, Portfolio and Asset Pricing. XII, 200 pages. 2002.

Vol. 515: G. Heisig, Planning Stability in Material Requirements Planning System. XII, 264 pages. 2002.

Vol. 516: B. Schmid, Pricing Credit Linked Financial Instruments. X, 246 pages. 2002.

Vol. 517: H. I. Meinhardt, Cooperative Decision Making in Common Pool Situations. VIII, 205 pages. 2002.

Vol. 518: S. Napel, Bilateral Bargaining. VIII, 188 pages. 2002.

Vol. 519: A. Klose, G. Speranza, L. N. Van Wassenhove (Eds.), Quantitative Approaches to Distribution Logistics and Supply Chain Management. XIII, 421 pages. 2002.

Vol. 520: B. Glaser, Efficiency versus Sustainability in Dynamic Decision Making. IX, 252 pages. 2002.

Vol. 521: R. Cowan, N. Jonard (Eds.), Heterogenous Agents, Interactions and Economic Performance. XIV, 339 pages. 2003.

Vol. 522: C. Neff, Corporate Finance, Innovation, and Strategic Competition. IX, 218 pages. 2003.

Vol. 523: W.-B. Zhang, A Theory of Interregional Dynamics. XI, 231 pages. 2003.

Vol. 524: M. Frölich, Programme Evaluation and Treatment Choise. VIII, 191 pages. 2003.

Vol. 525: S. Spinler, Capacity Reservation for Capital-Intensive Technologies. XVI, 139 pages. 2003.

Vol. 526: C. F. Daganzo, A Theory of Supply Chains. VIII, 123 pages. 2003.

Vol. 527: C. E. Metz, Information Dissemination in Currency Crises. XI, 231 pages. 2003.

Vol. 528: R. Stolletz, Performance Analysis and Optimization of Inbound Call Centers. X, 219 pages. 2003.

Vol. 529: W. Krabs, S. W. Pickl, Analysis, Controllability and Optimization of Time-Discrete Systems and Dynamical Games. XII, 187 pages. 2003.

Vol. 530: R. Wapler, Unemployment, Market Structure and Growth. XXVII, 207 pages. 2003.

Vol. 531: M. Gallegati, A. Kirman, M. Marsili (Eds.), The Complex Dynamics of Economic Interaction. XV, 402 pages, 2004.

Vol. 532: K. Marti, Y. Ermoliev, G. Pflug (Eds.), Dynamic Stochastic Optimization. VIII, 336 pages. 2004.

Vol. 533: G. Dudek, Collaborative Planning in Supply Chains. X, 234 pages. 2004.

Vol. 534: M. Runkel, Environmental and Resource Policy for Consumer Durables. X, 197 pages. 2004.

Vol. 535: X. Gandibleux, M. Sevaux, K. Sörensen, V. T'kindt (Eds.), Metaheuristics for Multiobjective Optimisation. IX, 249 pages. 2004.

Vol. 536: R. Brüggemann, Model Reduction Methods for Vector Autoregressive Processes. X, 218 pages. 2004.

Vol. 537: A. Esser, Pricing in (In)Complete Markets. XI, 122 pages, 2004.

Vol. 538: S. Kokot, The Econometrics of Sequential Trade Models. XI, 193 pages. 2004.

Vol. 539: N. Hautsch, Modelling Irregularly Spaced Financial Data. XII, 291 pages. 2004.

Vol. 540: H. Kraft, Optimal Portfolios with Stochastic Interest Rates and Defaultable Assets. X, 173 pages. 2004.

Vol. 541: G.-y. Chen, X. Huang, X. Yang, Vector Optimization. X, 306 pages. 2005.

Vol. 542: J. Lingens, Union Wage Bargaining and Economic Growth. XIII, 199 pages. 2004.

Vol. 543: C. Benkert, Default Risk in Bond and Credit Derivatives Markets. IX, 135 pages. 2004.

Vol. 544: B. Fleischmann, A. Klose, Distribution Logistics. X, 284 pages. 2004.

Vol. 545: R. Hafner, Stochastic Implied Volatility. XI, 229 pages. 2004.

Vol. 546: D. Quadt, Lot-Sizing and Scheduling for Flexible Flow Lines. XVIII, 227 pages. 2004.

Vol. 547: M. Wildi, Signal Extraction. XI, 279 pages. 2005.

Vol. 548: D. Kuhn, Generalized Bounds for Convex Multistage Stochastic Programs. XI, 190 pages. 2005.

Vol. 549: G. N. Krieg, Kanban-Controlled Manufacturing Systems. IX, 236 pages. 2005.

Vol. 550: T. Lux, S. Reitz, E. Samanidou, Nonlinear Dynamics and Heterogeneous Interacting Agents. XIII, 327 pages. 2005.

Vol. 551: J. Leskow, M. Puchet Anyul, L. F. Punzo, New Tools of Economic Dynamics. XIX, 392 pages. 2005.

Vol. 552: C. Suerie, Time Continuity in Discrete Time Models. XVIII, 229 pages. 2005.

Vol. 553: B. Mönch, Strategic Trading in Illiquid Markets. XIII, 116 pages. 2005.

Vol. 554: R. Foellmi, Consumption Structure and Macroeconomics. IX, 152 pages. 2005.

Vol. 555: J. Wenzelburger, Learning in Economic Systems with Expectations Feedback (planned) 2005.

Vol. 556: R. Branzei, D. Dimitrov, S. Tijs, Models in Cooperative Game Theory. VIII, 135 pages. 2005.

Vol. 557: S. Barbaro, Equity and Efficiency Considerations of Public Higer Education. XII, 128 pages. 2005.

Vol. 558: M. Faliva, M. G. Zoia, Topics in Dynamic Model Analysis. X, 144 pages. 2005.

Vol. 559: M. Schulmerich, Real Options Valuation. XVI, 357 pages. 2005.

Vol. 560: A. von Schemde, Index and Stability in Bimatrix Games. X, 151 pages. 2005.

Vol. 561: H. Bobzin, Principles of Network Economics. XX, 390 pages. 2006.

Vol. 562: T. Langenberg, Standardization and Expectations. IX, 132 pages. 2006.

Vol. 563: A. Seeger (Ed.), Recent Advances in Optimization. XI, 455 pages. 2006.

Vol. 564: P. Mathieu, B. Beaufils, O. Brandouy (Eds.), Artificial Economics. XIII, 237 pages. 2005.

Vol. 565: W. Lemke, Term Structure Modeling and Estimation in a State Space Framework. IX, 224 pages. 2006.

Vol. 566: M. Genser, A Structural Framework for the Pricing of Corporate Securities. XIX, 176 pages. 2006.

Vol. 567: A. Namatame, T. Kaizouji, Y. Aruga (Eds.), The Complex Networks of Economic Interactions. XI, 343 pages. 2006.

Vol. 568: M. Caliendo, Microeconometric Evaluation of Labour Market Policies. XVII, 258 pages. 2006.

Vol. 569: L. Neubecker, Strategic Competition in Oligopolies with Fluctuating Demand. IX, 233 pages. 2006.

Vol. 570: J. Woo, The Political Economy of Fiscal Policy. X, 169 pages. 2006.

Vol. 571: T. Herwig, Market-Conform Valuation of Options. VIII, 104 pages. 2006.

Vol. 572: M. F. Jäkel, Pensionomics. XII, 316 pages. 2006

Vol. 573: J. Emami Namini, International Trade and Multinational Activity, X, 159 pages, 2006.

Vol. 574: R. Kleber, Dynamic Inventory Management in Reverse Logisnes, XII, 181 pages, 2006.

Vol. 575: R. Hellermann, Capacity Options for Revenue Management, XV, 199 pages, 2006.

Vol. 576: J. Zajac, Economics Dynamics, Information and Equilibnum, X, 284 pages, 2006.

Vol. 577: K. Rudolph, Bargaining Power Effects in Financial Contracting, XVIII, 330 pages, 2006.

Vol. 578: J. Kühn, Optimal Risk-Return Trade-Offs of Commercial Banks, IX, 149 pages, 2006.

Vol. 579: D. Sondermann, Introduction to Stochastic Calculus for Finance, X, 136 pages, 2006.

Vol. 580: S. Seifert, Posted Price Offers in Internet Auction Markets, IX, 186 pages, 2006.

Vol. 581: K. Marti; Y. Ermoliev; M. Makowsk; G. Pflug (Eds.), Coping with Uncertainty, XIII, 330 pages, 2006 (planned).

Vol. 582: J. Andritzky, Sovereign Default Risks Valuation: Implications of Debt Crises and Bond Restructurings. VIII, 251 pages, 2006 (planned).

Vol. 583: I.V. Konnov, D.T. Luc, A.M. Rubinov (Eds.), Generalized Convexity and Related Topics, IX, 469 pages, 2006 (planned).

Vol. 584: C. Bruun, Adances in Artificial Economics: The Economy as a Complex Dynamic System. XVI, 296 pages, 2006.

Vol. 585: R. Pope, J. Leitner, U. Leopold-Wildburger, The Knowledge Ahead Approach to Risk, XVI, 218 pages, 2007 (planned).

Vol. 586: B. Lebreton, Strategic Closed-Loop Supply Chain Management. X, 150 pages, 2007 (planned).

Printing: Krips bv, Meppel
Binding: Stürtz, Würzburg